Official Certified SOLIDWORKS Professional (CSWP) Certification Guide

SOLIDWORKS 2015 – SOLIDWORKS 2017

David C. Planchard

CSWP & SOLIDWORKS Accredited Educator

Publications

SDC Publications
P.O. Box 1334
Mission, KS 66222
913-262-2664
www.SDCpublications.com
Publisher: Stephen Schroff

ISBN-13: 978-1-63057-071-2
ISBN-10: 1-63057-071-0

Printed and bound in the United States of America.

INTRODUCTION

The **Official Certified SOLIDWORKS® Professional (CSWP) Certification Guide, Version 4: 2017, 2016, 2015** is written to assist the SOLIDWORKS user to take and pass the CSWP CORE exam.

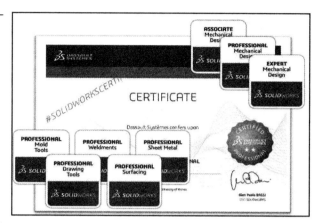

The book is organized into CSWP CORE segments: 1, 2 and 3. It is written to help you pass the three sections of the CSWP CORE exam.

It provides CSWP exam tips, hints and information on sample questions and categories that are aligned with each segment of the CSWP CORE exam.

Redeem the code on the inside cover of the book. Download the videos and model folders to a local hard drive. View the provided videos and models to enhance the user experience.

📁 CSWP model folder 2015
📁 CSWP model folder 2016
📁 CSWP model folder 2017

Work directly from your hard drive on the segment chapters in the book. SOLIDWORKS model files (initial and final) for 2015, 2016 and 2017 are provided.

The CSWP Certification exam is offered in three separate segments. The book covers the three segments of the CSWP CORE exam. At this time for SOLIDWORKS World, there is a single three hour CSWP exam only for registered users.

📁 Segment 1 Final
📁 Segment 1 Initial
📁 Segment 2 Final
📁 Segment 2 Initial
📁 Segment 3 Final
📁 Segment 3 Initial

The first segment is not a prerequisite for the second and the second segment is not a prerequisite for the third. You can take the segments in any order.

Each segment covers a different set of disciplines. All segments are timed. You will be tested on data found in the Mass Properties section of SOLIDWORKS. It is important to be familiar with accessing Mass Properties and interpreting them correctly.

The first segment is 90 minutes with six (6) questions. The format is either multiple choice or single fill in the blank. The first question in each segment is typically in a multiple choice format. A total score of 85 out of 115 or better is required to pass the first segment.

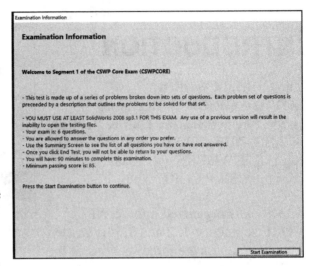

Actual CSWP exam format

You need the exact answer (within 1% of the stated value in the multiple choice section) before you should move on to the next question which is in a single answer format.

If you do not have the exact answer to the first question, you will most likely fail the following questions. This is crucial as there is no partial credit.

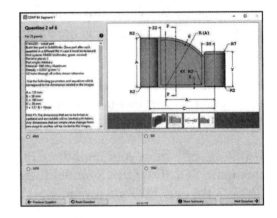

Actual CSWP exam format

In the first segment, you will create a part and modify various dimensions. A question is presented to you in multiple steps and you need to obtain the correct answer at each step to get the question correct. There is no partial credit in the exam.

The Examination Information dialog box states that you are allowed to answer the questions in any order that you prefer. Is this true?

Yes and No! Segment 1 consists of 6 questions. In those six (6) questions there are three major stages to be modeled: Stage 1 (question 1), Stage 2 (question 4) and Stage 3 (question 5). Questions 2, 3 and 6 then ask the user to make a modification to those stages.

Strategically it is best if a user does questions 1, 4 and 5 to model the major stages and compare their Mass Properties to the multiple choice answer. You should be within 1% of the stated value in the multiple choice section before you go back to questions 2, 3 and 6 (single fill in the blank format) to make the changes.

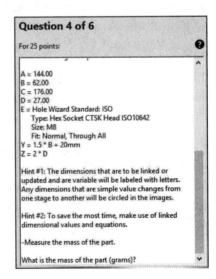

The second segment of the CSWP CORE exam is 40 minutes with nine (9) questions (11 total but two are instructional pages) divided into two categories. The format is either multiple choice or single fill in the blank. The first question in this segment is typically in a multiple choice format.

The second segment address:

- Creating configurations from other configurations

- Modifying configurations using a design table

- Obtaining Mass Properties of various modified configurations

- Modifying features and sketches of an existing part

- Recovering from rebuild errors

A total score of 115 out of 155 or better is required to pass the second segment.

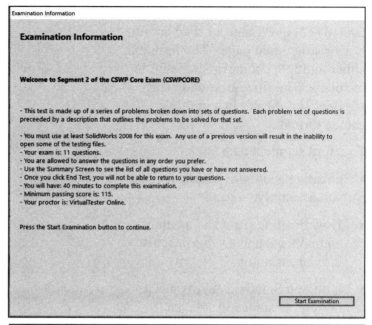

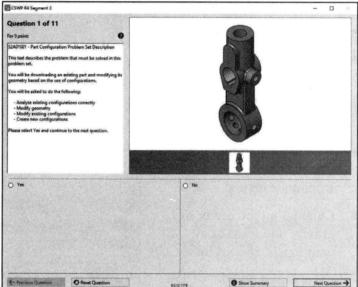

Actual CSWP exam format

The third segment is 80 minutes with twelve (12) questions (13 total but one is an instructional page). The format is either multiple choice or single fill in the blank. The first question in each segment is typically in a multiple choice format.

The third segment addresses:

- Creating a simple component and sub-assembly.

- Downloading parts and assembling components and a sub-assembly into an assembly.

- Utilizing Standard & Advanced (Width, Distance, Angle, etc.) mates.

- Applying Rigid and Flexible states.

- Employing the Interference Detection tool and Collision tool.

- Creating and using a Coordinate System.

- Modifying and replacing components and recovering from mate errors.

- Calculating the center of gravity and mass of a component and assembly.

A total score of 140 out of 190 or better is required to pass the third segment.

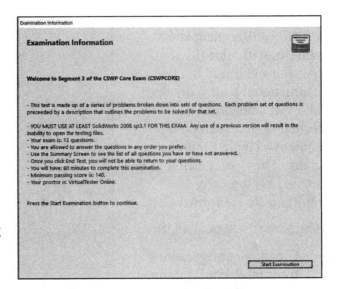

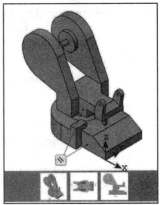

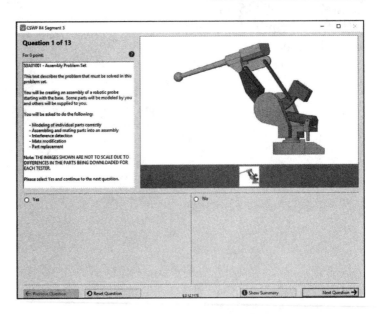

Actual CSWP exam format

Once you pass *all three segments*, you will receive the following email for your CSWP CORE certification in a day or two. Click on the SOLIDWORKS Certification Center hyperlink to log in, activate and view your certificate.

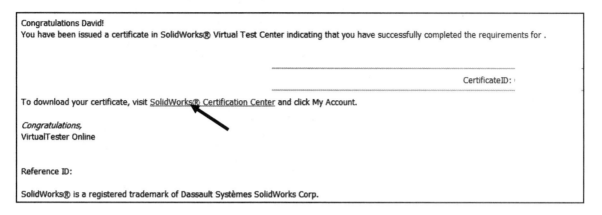

Congratulations David!
You have been issued a certificate in SolidWorks® Virtual Test Center indicating that you have successfully completed the requirements for .

CertificateID:

To download your certificate, visit SolidWorks® Certification Center and click My Account.

Congratulations,
VirtualTester Online

Reference ID:

SolidWorks® is a registered trademark of Dassault Systèmes SolidWorks Corp.

Valid Certificates						
	Certificate ID	Certificate	Issued By	Issued On		
	C-B9f	Segment 2 of Certified SolidWorks Pr...	Avelino Rochino, SolidWorks Corporation	02/14/2011		
	C-UPi	Segment 1 of Certified SolidWorks Pr...	Avelino Rochino, SolidWorks Corporation	02/13/2011		
	C-HV.	Segment 3 of Certified SolidWorks Pr...	Avelino Rochino, SolidWorks Corporation	03/06/2011		
☆	C-VU	Certified SolidWorks Professional - C...	VirtualTester Online, Tangix Design & Devel...	03/06/2011		

🔆 If you fail a segment of the exam, you need to wait 14 days until you can retake that same segment. In that time, you can take another segment.

During the exam, SOLIDWORKS provides the ability to click on a view to obtain additional details and dimensions.

🔆 FeatureManager names were changed through various revisions of SOLIDWORKS. Example: Extrude1 vs. Boss-Extrude1. These changes do not affect the models or answers in this book.

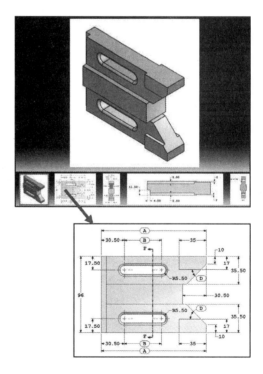

Goals

The primary goal is not only to help you pass the CSWP CORE exam, but also to ensure that you understand and comprehend the concepts and implementation details of the process.

The second goal is to provide the most comprehensive coverage of CSWP CORE exam related topics available, without too much coverage of topics not on the exam.

The third and ultimate goal is to get you from where you are today to the point that you can confidently pass all three segments of the CSWP CORE exam. You must work quickly and accurately.

CSWP Audience

The intended audience for this book and the CSWP exam is a person who has passed the CSWA exam and who has 8 or more months of SOLIDWORKS training and usage.

The CSWA exam is not a prerequisite for the CSWP exam if you are a commercial user in industry. For students that take the CSWP exam through their school, you must first pass the CSWA exam.

About the Author

David Planchard is the founder of D&M Education LLC. Before starting D&M Education, he spent over 27 years in industry and academia holding various engineering, marketing, and teaching positions. He holds five U.S. patents. He has published and authored numerous papers on Machine Design, Product Design, Mechanics of Materials, and Solid Modeling. He is an active member of the SOLIDWORKS Users Group and the American Society of Engineering Education (ASEE). David holds a BSME, MSM with the following professional certifications: CCAI, CCNP, CSDA, CSWSA-FEA, CSWP, CSWP-DRWT and SOLIDWORKS Accredited Educator. David is a SOLIDWORKS Solution Partner, an Adjunct Faculty member and the SAE advisor at Worcester Polytechnic Institute in the Mechanical Engineering department. In 2012, David's senior Major Qualifying Project team (senior capstone) won first place in the Mechanical Engineering department at WPI. In 2014, 2015 and 2016, David's senior Major Qualifying Project team won the Provost award in Mechanical Engineering for design excellence.

David Planchard is the author of the following books:

- **SOLIDWORKS® 2017 Reference Guide with video instruction**, 2016, 2015, 2014, 2013, 2012, 2011, 2010 and 2009

- **Engineering Design with SOLIDWORKS® 2017 and video instruction**, 2016, 2015, 2014, 2013, 2012, 2011, 2010, 2009, 2008, 2007, 2006, 2005, 2004, and 2003

- **Engineering Graphics with SOLIDWORKS® 2017 and video instruction**, 2016, 2015, 2014, 2013, 2012 and 2011

- **SOLIDWORKS® 2017 in 5 Hours with video instruction**, 2016, 2015, 2014

- **SOLIDWORKS® 2017 Tutorial with video instruction**, 2016, 2015, 2014, 2013, 2012, 2011, 2010, 2009, 2008, 2007, 2006, 2005, 2004 and 2003

- **Drawing and Detailing with SOLIDWORKS® 2014**, 2012, 2010, 2009, 2008, 2007, 2006, 2005, 2004, 2003, and 2002

- **Official Certified SOLIDWORKS® Professional (CSWP) Certification Guide with video instruction, Version 4: 2015 - 2017**, Version 3: 2012 - 2014, Version 2: 2012 - 2013, Version 1: 2010 - 2010

- **Official Guide to Certified SOLIDWORKS® Associate Exams: CSWA, CSDA, CSWSA-FEA Version 3: 2015 - 2017**, Version 2: 2012 - 2015, Version 1: 2012 - 2013

- **Assembly Modeling with SOLIDWORKS® 2012**, 2010, 2008, 2006, 2005-2004, 2003 and 2001Plus

- **Applications in Sheet Metal Using Pro/SHEETMETAL & Pro/ENGINEER**

Acknowledgements

Writing this book was a substantial effort that would not have been possible without the help and support of my loving family and of my professional colleagues. I would like to thank Professor John M. Sullivan Jr., Professor Jack Hall and the community of scholars at Worcester Polytechnic Institute who have enhanced my life, my knowledge and helped to shape the approach and content to this text.

The author is greatly indebted to my colleagues from Dassault Systèmes SOLIDWORKS Corporation for their help and continuous support: Avelino Rochino and Mike Puckett.

Thanks also to Professor Richard L. Roberts of Wentworth Institute of Technology, Professor Dennis Hance of Wright State University, Professor Jason Durfess of Eastern Washington University and Professor Aaron Schellenberg of Brigham Young University - Idaho who provided vision and invaluable suggestions.

SOLIDWORKS certification has enhanced my skills and knowledge and that of my students. Thank you to Ian Matthew Jutras (CSWE), who is a technical contributor and the creator of the videos, and Stephanie Planchard, technical procedure consultant.

Contact the Author

We realize that keeping software application books current is imperative to our customers. We value the hundreds of professors, students, designers, and engineers that have provided us input to enhance the book. Please contact me directly with any comments, questions or suggestions on this book or any of our other SOLIDWORKS books at dplanchard@msn.com or planchard@wpi.edu.

Note to Instructors

Please contact the publisher **www.SDCpublications.com** for classroom support materials (.ppt presentations, labs and more) and the Instructor's Guide with model solutions and tips that support the usage of this text in a classroom environment.

Trademarks, Disclaimer and Copyrighted Material

SOLIDWORKS®, eDrawings®, SOLIDWORKS Simulation®, SOLIDWORKS Flow Simulation, and SOLIDWORKS Sustainability are a registered trademark of Dassault Systèmes SOLIDWORKS Corporation in the United States and other countries; certain images of the models in this publication courtesy of Dassault Systèmes SOLIDWORKS Corporation.

Microsoft Windows®, Microsoft Office® and its family of products are registered trademarks of the Microsoft Corporation. Other software applications and parts described in this book are trademarks or registered trademarks of their respective owners.

The publisher and the author make no representations or warranties with respect to the accuracy or completeness of the contents of this work and specifically disclaim all warranties, including without limitation warranties of fitness for a particular purpose. No warranty may be created or extended by sales or promotional materials. Dimensions of parts are modified for illustration purposes. Every effort is made to provide an accurate text. The author and the manufacturers shall not be held liable for any parts, components, assemblies or drawings developed or designed with this book or any responsibility for inaccuracies that appear in the book. Web and company information was valid at the time of this printing.

The Y14 ASME Engineering Drawing and Related Documentation Publications utilized in this text are as follows: ASME Y14.1 1995, ASME Y14.2M-1992 (R1998), ASME Y14.3M-1994 (R1999), ASME Y14.41-2003, ASME Y14.5-1982, ASME Y14.5-1999, and ASME B4.2. Note: By permission of The American Society of Mechanical Engineers, Codes and Standards, New York, NY, USA. All rights reserved.

Additional information references the American Welding Society, AWS 2.4:1997 Standard Symbols for Welding, Braising, and Non-Destructive Examinations, Miami, Florida, USA.

References

- SOLIDWORKS Help Topics and What's New, SOLIDWORKS Corporation, 2017.

- Beers & Johnson, Vector Mechanics for Engineers, 6th ed. McGraw Hill, Boston, MA.

- Gradin, Hartley, Fundamentals of the Finite Element Method, Macmillan, NY 1986.

- Hibbler, R.C, Engineering Mechanics Statics and Dynamics, 8th ed, Prentice Hall.

- Jensen & Helsel, Engineering Drawing and Design, Glencoe, 1990.

- Lockhart & Johnson, Engineering Design Communications, Addison Wesley, 1999.

- Olivo C., Payne, Olivo, T, Basic Blueprint Reading and Sketching, Delmar 1988.

- Walker, James, Machining Fundamentals, Goodheart Wilcox, 1999.

- 80/20 Product Manual, 80/20, Inc., Columbia City, IN, 2012.

- Ticona Designing with Plastics - The Fundamentals, Summit, NJ, 2009.

- SMC Corporation of America, Product Manuals, Indiana, USA, 2012.

- Emerson-EPT Bearing Product Manuals and Gear Product Manuals, Emerson Power Transmission Corporation, Ithaca, NY, 2009.

- Emhart - A Black and Decker Company, On-line catalog, Hartford, CT, 2012.

During the initial SOLIDWORKS installation, you are requested to select either the ISO or ANSI drafting standard. ISO is typically a European drafting standard and uses First Angle Projection. The book is written using the ANSI (US) overall drafting standard and Third Angle Projection for drawings.

Redeem the code on the inside cover of the book. View the provided models to enhance the user experience.

TABLE OF CONTENTS

About the Book

The intended audience for the book is a person who has passed the CSWA exam and who has 8 or more months of SOLIDWORKS training and usage.

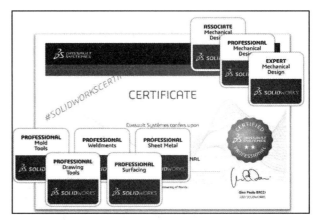

The book is organized into CSWP CORE segments: 1, 2 and 3.

It is written to help you pass the three sections of the CSWP CORE exam.

Each chapter is focused on a segment of the CSWP CORE exam. This is not a step-by-step book.

It provides CSWP exam tips, hints and information on sample questions and categories that are aligned with the exam.

It is not written to teach a new user SOLIDWORKS.

Redeem the code on the inside cover of the book. Download the videos and CSWP model folder.

CSWP model folder 2015
CSWP model folder 2016
CSWP model folder 2017

Work directly from your hard drive on the segment chapters in the book. SOLIDWORKS model files (initial and final) for 2015, 2016 and 2017 are provided.

Segment 1 Final
Segment 1 Initial
Segment 2 Final
Segment 2 Initial
Segment 3 Final
Segment 3 Initial

Utilize the provided segment model folders to follow along while using the book.

TESTING TIPS

I have collected various testing tips over the years from my colleagues and friends on the SOLIDWORKS Certification exams. Below is a list that may help you prepare and pass the CSWP exam.

1. It is an exam. It is timed. There is no partial credit. Be precise in your work and entering your answers with the correct amount of decimal places. Were you perfect in the sample CSWP exam or on the examples in this book? If not, why? Correct the mistakes before taking the real exam; you will be glad you did.

2. The sample CSWP exam only covers Segment 1 of the CSWP CORE exam. Time yourself on the practice exam. You should be able to finish the sample Segment 1 exam in approximately 60 - 75 minutes.

3. Read up on the contents of the exam that you are not familiar with (collision detection, interference detection, Advanced mates, measure tool, design tables, equations, coordinate locations, replace components, etc.) before you take any segment of the CSWP CORE exam.

4. You will be tested on data found in the Mass Properties section of SOLIDWORKS. It is important to be familiar with accessing Mass Properties and interpreting them correctly.

5. The first question in each segment is usually in a multiple choice format. You need the exact answer (within 1% of the stated value) before you move on to the next question (fill in the blank). If you don't find your answer (within 1%) in the multiple choice single answer format section, recheck your solid model for precision and accuracy.

6. SOLIDWORKS Mass Properties calculates the center of mass for a model. At every instant of time, there is a unique location (x, y, z) in space that is the average position of the systems mass. The CSWP exam asks for center of gravity. For the purpose of calculating the center of mass and center of gravity near to earth or on earth, you can assume that the center of mass and the center of gravity are the same.

7. In Segment 1 of the CSWP CORE exam use Link Values for variables A thru E. Use Equations for X & Y. This will save you time. With Link Values, modify the dimensions, not the equations.

8. Create a directory and file structure to save your model during the exam. Create a millimeter, 2 decimal place, part template and assembly template.

9. Take the test on a system you are familiar with. Don't customize your system right before you take the exam.

10. Read all questions before beginning. Pay attention to material changes, origin location, dimensional changes. Note which values indicated by a letter are the ones that will change.

11. Rely only on dimensions and provided information. Do NOT rely on the image provided as a template. Dimensions are changed from test to test so the image used will not be to scale (this is noted in the exam).

12. SOLIDWORKS displays a circle, ellipse or a square around the areas and features that require modification from the original part.

13. Notice where the dimensions are referenced in the drawing views. If the dimensions are referenced from the lower right-hand corner, this is where you should begin with the origin for your Base Sketch (Sketch1).

14. If you use faces versus a dimension, this can help when you need to create a design change to maintain the design intent of the part.

15. Suppress features before you delete them. This will inform you if there are any rebuild or feature errors during modification.

16. A question will be presented to you in multiple steps and you need to get each step correct to get the final question correct. There is no partial credit on the exam.

17. Always enter the needed decimal places in the answer field even if 0's. Example: 120.00.

18. Always confirm that your math is correct. Use the Measure tool.

19. As a general rule, insert relations before dimensions in a sketch.

20. Fillets are modified often. Use caution when taking the exam.

21. Utilize the split bar to work between the FeatureManager and the ConfigurationManager in Segment 2 of the CSWP CORE exam.

22. In Segment 3 of the CSWP CORE exam, the first component in the assembly should be fixed to the origin or fully defined.

23. You may be required to measure angles between flat surfaces. Be certain to understand the direction, complement or supplement of the required angle.

24. Confirm the position of components inside an assembly when using the Replace Components tool. Apply the Change Transparency tool or a section view to confirm mated component location.

25. SOLIDWORKS will provide a part to create that is not orientated correctly in the assembly. Knowledge of component orientation and changing orientation in an assembly is required along with creating a coordinate system.

26. Save your work frequently and rename saved parts in all segments of the exam.

27. Imported components have imported geometry. Select **NO** on Feature Recognition. Import the geometry as quickly as possible.

28. You must use SOLIDWORKS 2008 SP3.1 or a newer version to take all three segments of the CSWP CORE exam.

29. Relax. Exam anxiety can be a killer. Take a deep breath and enjoy.

There are numerous ways to build the models in this book. A goal is to display different design intents and techniques.

CHAPTER 1 - SEGMENT 1 OF THE CSWP CORE EXAM

Introduction

DS SOLIDWORKS Corp. offers various stages of certification. Each stage represents increasing levels of expertise in 3D CAD design: *Certified SOLIDWORKS Associate CSWA, Certified SOLIDWORKS Professional CSWP and Certified SOLIDWORKS Expert CSWE* along with specialty fields in Drawings, Simulation, Sheet Metal, Surfacing and more.

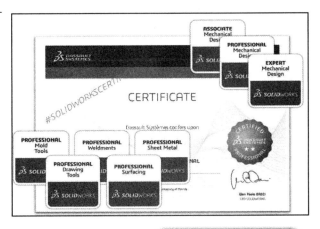

The CSWP Certification exam is offered in three separate segments. The book covers the three separate segments of the CSWP CORE exam.

The first segment is not a prerequisite for the second, and the second segment is not a prerequisite for the third. You can take the segments in any order.

Each segment covers a different set of disciplines. This chapter addresses the first segment of the CSWP CORE exam. All segments are timed. You will be tested on data found in the Mass Properties section of SOLIDWORKS. It is important to be familiar with accessing Mass Properties and interpreting them correctly.

The first segment is 90 minutes with six (6) questions. The format is multiple choice and single fill in the blank.

The first question in each segment is typically in a multiple choice format. A total score of 85 out of 115 or better is required to pass the first segment.

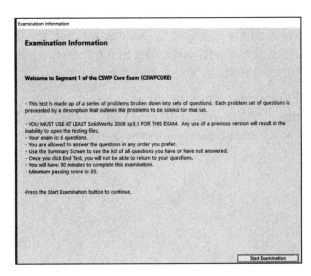

Actual CSWP exam format

You should have the exact answer (within 1% of the stated value in the multiple choice section) before you move to the next question. If you don't have the exact answer for the first question, you will most likely fail the following question. This is crucial as there is no partial credit.

In the first segment, you will create a part and modify various dimensions. A question is presented to you in multiple steps, and you need to obtain the correct answer at each step.

🔆 To save time, utilize global variables and equations.

The Examination Information dialog box states that you are allowed to answer the questions in any order. Is this true?

Yes and No! Segment 1 consists of six (6) questions. In those 6 questions there are three major stages to be modeled: Stage 1 (question 1), Stage 2 (question 4) and Stage 3 (question 5).

Questions 2, 3 and 6 then ask the user to make a modification to those stages.

Strategically it is best if a user does questions 1, 4 and 5 to model the major stages and compare the Mass Properties to the multiple choice answer. You should be within 1% of the stated value in the multiple choice section before you go back to questions 2, 3 and 6 (fill in the blank format) to make the changes.

Engineering drawing views with annotations are presented to you in all segments of the CSWP CORE exam. Review the next section.

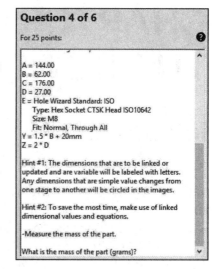

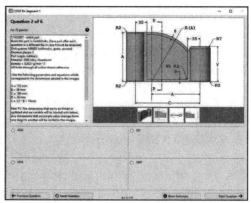

Actual CSWP exam format

Read and understand an Engineering document

What is an Engineering document? In SOLIDWORKS a part, assembly or drawing is referred to as a document. Each document is displayed in the Graphics window.

During the exam, each question will display an information table on the left side of the screen and drawing information on the right.

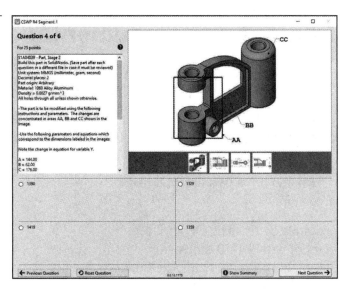

Actual CSWP exam format

Read the provided information and apply it to the drawing. Various values are provided in each question.

☼ If you do not find your answer (within 1%) in the multiple choice single answer format section - recheck your solid model for precision and accuracy.

☼ SOLIDWORKS Mass Properties calculates the center of mass for every model. At every instant of time, there is a unique location (x, y, z) in space that is the average position of the mass in the system. The CSWP exam asks for center of gravity. For the purpose of calculating the center of mass and center of gravity near to earth or on earth, you can assume that the center of mass and the center of gravity are the same.

☼ SOLIDWORKS views present illustrations that are not proportional to the given dimensions.

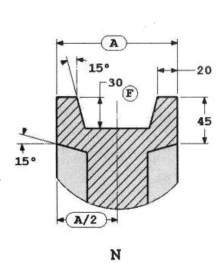

N

Engineering Documentation Practices

2D drawing views are displayed in the CSWP exam. The ability to interpret a 2D drawing view is required.

Example 1: *8X Ø.19 EQ. SP*. Eight holes with a .19in. diameter is required that are equally (.55in.) spaced.

Example 2: *R2.50 TYP*. Typical radius of 2.50. The dimension has a two decimal place precision.

Example 3: ⊽. The Depth/Deep ⊽ symbol with a 1.50 dimension associated with the hole. The hole Ø.562 has a three decimal place precision.

Example 4: *A+40*. A is provided to you on the exam. 44mm + A.

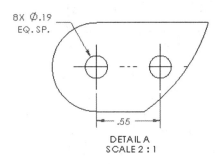

N is a Detail view of the M-M Section view.

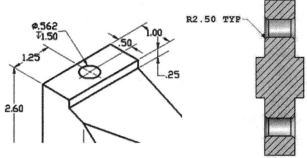

- Example 5: *ØB*. Diameter of B. B is provided to you on the exam.

- Example 6: ⌰. Parallelism.

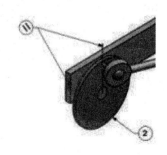

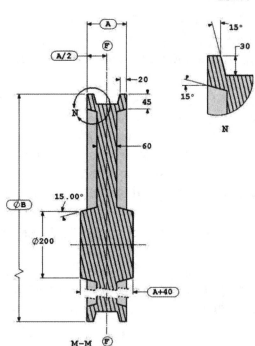

Build a part from a detailed illustration

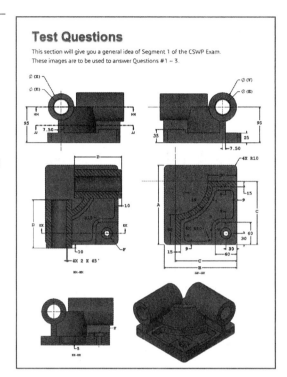

Test Questions

This section will give you a general idea of Segment 1 of the CSWP Exam. These images are to be used to answer Questions #1 – 3.

Segment 1 of the CSWP CORE exam - First question

Segment 1 of the CSWP CORE exam is one of three sections. In it, you are asked to create and modify a single part. Segment 2 and 3 of the exam requires you to download part and component files using the testing client. The question setup in segment 1 is similar to the sample test posted by SOLIDWORKS.

Below is the needed information and steps to correctly create the provided model in the sample exam.

 View the provided pdf file if needed for this segment.

 Utilize segment model folders to follow along while using the book.

Provided information:

Initial part - Stage 1: Build this part in SOLIDWORKS.

Unit system: MMGS (millimeter, gram, second)

Decimal places: 2

Part origin: Arbitrary

Material: Alloy Steel

Density: 0.0077 g/mm^3

All holes through all unless shown otherwise.

Use the following parameters and equations which correspond to the dimensions labeled in the images:

A = 213 mm

B = 200 mm

C = 170 mm

D = 130 mm

E = 41 mm

F = Hole Wizard Standard: ANSI Metric - Counterbore

> Type: Hex Bolt - ANSI B18.2.3.5M
>
> Size: M8
>
> Fit: Close
>
> Through Hole Diameter: 15.00 mm
>
> Counterbore Diameter: 30.00 mm
>
> Counterbore Depth: 10.00 mm
>
> End Condition: Through All

X = A/3

Y = B/3 + 10mm

Hint #1: The dimensions that are to be linked or updated and are variable will be labeled with letters. Any dimensions that are simple value changes from one stage to another will be circled in the images.

Hint #2: To save the most time, make use of linked dimensional values and equations.

Measure the mass of the part.

What is the mass of the part (grams)?

a) 14139.65

b) 14298.56

c) 15118.41

d) 14207.34

1. Stage 1 – Initial Part
Build this part in SOLIDWORKS.

Unit system: MMGS (millimeter, gram, second)
Decimal places: 2
Part origin: Arbitrary
Material: Alloy Steel
Density = 0.0077 g/mm^3
All holes through all unless shown otherwise

Use the following parameters and equations which correspond to the dimensions labeled in the images:

A = 213 mm
B = 200 mm
C = 170 mm
D = 130 mm
E = 41 mm
F = Hole Wizard Standard: Ansi Metric Counterbore
    Type: Hex Bolt – ANSI B18.2.3.5M
    Size: M8
    Fit: Close
    Through Hole Diameter: 15.00 mm
    Counterbore Diameter: 30.00 mm
    Counterbore Depth: 10.00 mm
    End Condition: Through All
X = A/3
Y = B/3 + 10mm

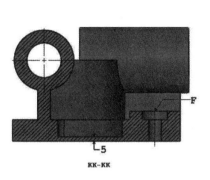 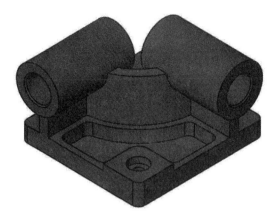

KK-KK

When you begin segment 1 of the CSWP CORE exam, you will be presented with a variety of drawings and parameters specified in the question. Take your time to first identify the drawing views, and to better understand the provided geometry that is needed.

When you create the initial part, think about using global variables, equations or design tables. This is crucial in this section of the book due to time constraints.

Note where the variables are (A, B, C, D & E) located in the provided drawing views. F is a Hole Wizard hole.

You will see anywhere between 5 to 6 variables and 2 equations in this segment of the exam.

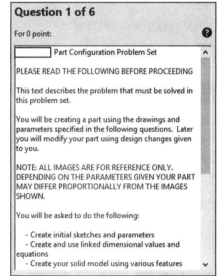

Question 1 of 6

For 0 point:

Part Configuration Problem Set

PLEASE READ THE FOLLOWING BEFORE PROCEEDING

This text describes the problem that must be solved in this problem set.

You will be creating a part using the drawings and parameters specified in the following questions. Later you will modify your part using design changes given to you.

NOTE: ALL IMAGES ARE FOR REFERENCE ONLY. DEPENDING ON THE PARAMETERS GIVEN YOUR PART MAY DIFFER PROPORTIONALLY FROM THE IMAGES SHOWN.

You will be asked to do the following:

- Create initial sketches and parameters
- Create and use linked dimensional values and equations
- Create your solid model using various features

Actual CSWP exam format

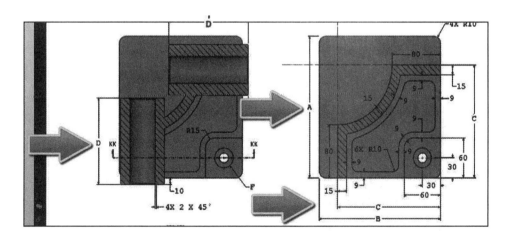

Observe where the dimensions are referenced. In the illustrated example, the dimensions are referenced from the lower right-hand corner. This is where you should begin with the origin of the rectangle for the Base Sketch (Sketch1) in this example.

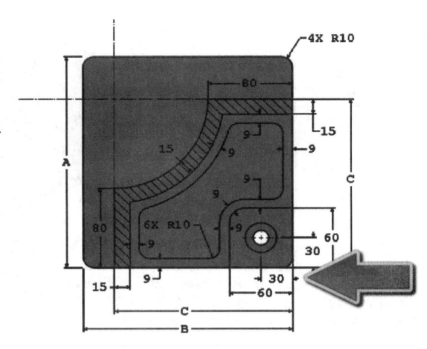

Most questions ask for a center of gravity or mass in grams. For the former it is crucial that the model is oriented exactly as illustrated. For the latter, it does not matter as much, but do not take the chance.

SOLIDWORKS previously used a tool called the Link Value, but this is no longer used. For users familiar with older versions of SOLIDWORKS (2012 or older), the Link Value was what we now think of as a Global Variable.

The text in this section is focused on using Global Variables with the equation drop-down menus in the Modify dialog box and in the Depth box in the PropertyManager.

A Global Variable is just a name assigned to a dimension, a reference measurement, or an entire equation. Global Variables are assigned in the Equations dialog box or in the Modify dimension box as simply the variable name equaling an expression of a value.

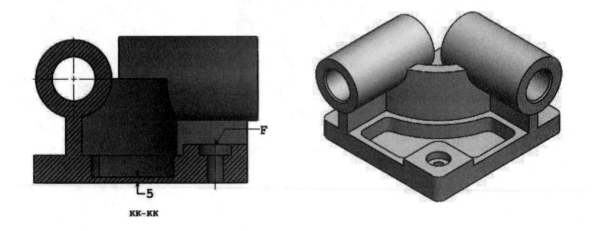

Be aware of what is symmetrical and what is different. The two cylinders look the same, but they are not. They have different diameters. It would not make sense to create a pattern or to mirror the two.

Look at the provided variables. Ask yourself, what requires a Global Variable in the Equation folder to address speedy part modification in a timed exam?

Understand the provided Hole type.

Based on the provided information, create an Equation folder and input the variables.

SOLIDWORKS displays a circle, ellipse or a square around the areas and features that require modification from the original part.

Remember, the purpose of this book is not to educate a new or intermediate user on SOLIDWORKS, but to inform them on the types of questions, layout and what to expect when taking the three segments of the CSWP CORE exam.

In this section, address part modification through an Equation folder using Global Variables, Equations or a design table.

Perform the procedure that you are the most comfortable with.

☀ The Modify dialog box accepts equations. You can also use it to create on-the fly Global Variables. To start an equation in the Modify box, start by replacing the numeric value with an = sign in the value box. When you do this, you will see a drop-down menu for functions and file properties. Use the dimension entry boxes in the PropertyManager.

A = 213 mm
B = 200 mm
C = 170 mm
D = 130 mm
E = 41 mm
F = Hole Wizard Standard: Ansi Metric Counterbore
  Type: Hex Bolt – ANSI B18.2.3.5M
  Size: M8
  Fit: Close
  Through Hole Diameter: 15.00 mm
  Counterbore Diameter: 30.00 mm
  Counterbore Depth: 10.00 mm
  End Condition: Through All
X = A/3
Y = B/3 + 10mm

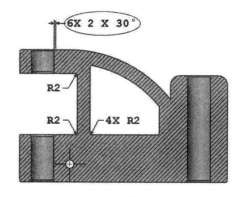

W-W

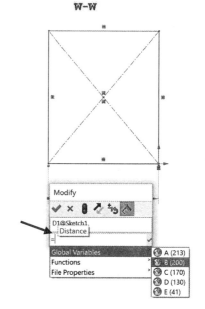

Let's begin.

1. **Create** a folder to save your models.

2. **Create** a new part.

3. **Set** document properties (drafting standard, units and precision) for the model.

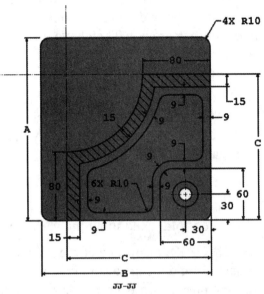

Start with setting the Global Variables. These are the provided variables:

A = 213 mm

B = 200 mm

C = 170 mm

D = 130 mm

E = 41 mm

4. **Display** the Equation folder in the FeatureManager.

5. **Display** the Equations, Global Variables, and Dimension dialog box.

6. **Enter** the five Global Variables (A, B, C, D & E) as illustrated.

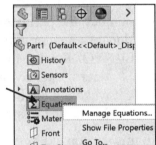

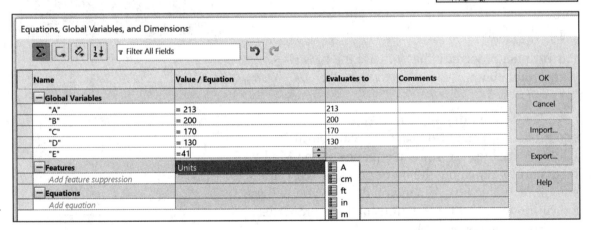

7. **Exit** the dialog box.

View the created Global Variables.

8. **Expand** the Equation folder in the FeatureManager.

Next, create the Base Sketch.

9. **Create** Sketch1. Select the Top Plane as the Sketch plane. Sketch1 is the profile for the Extruded Base (Boss-Extrude1) feature. Apply the Corner Rectangle Sketch tool. Click the origin and a position in the upper left section of the Graphics window. Most of the dimensions in the provided drawing view are referenced from this location.

10. **Insert** the horizontal dimension using the Global Variable B(200) from the Modify dialog box. Enter an equal sign (=) first in the Modify dialog box.

11. **Enter** B for Dimension Text. This will help you keep track of the variables.

12. **Insert** the vertical dimension using the Global Variable A(213) from the Modify dialog box. Sketch1 is fully defined.

13. **Enter** A for Dimension Text.

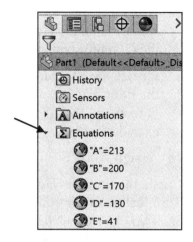

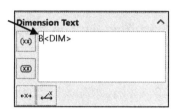

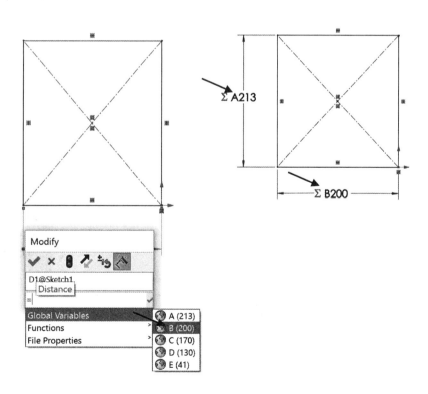

14. **Display** Primary values. Click View Dimension Names from the Heads-up toolbar. View the results in the Graphics window.

15. **Create** the Extruded Base feature. Boss-Extrude1 is the Base feature. Blind is the default End Condition in Direction 1. Depth = 25mm. Note the direction of the extrude feature.

16. **Assign** Alloy Steel material to the part.

Create the Extrude-Thin feature. The Extrude-Thin feature is controlled by the variable C and two dimensions that are equal (80mm). The height of the feature from the bottom is 95mm.

Start from the Top face and subtract 25mm from 95mm to obtain the correct depth for the Thin-Extrude feature.

17. **Create** Sketch2. Select the Top face. Sketch2 is the profile for the Extrude-Thin feature. Utilize the Line Sketch tool to create a vertical and horizontal line. Utilize the 3 Point Arc Sketch tool to complete the sketch.

18. **Add** an Equal relation between the vertical and horizontal line of Sketch2.

19. **Insert** the 80mm dimension on the vertical line.

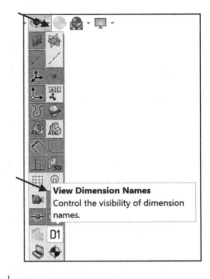

View Dimension Names
Control the visibility of dimension names.

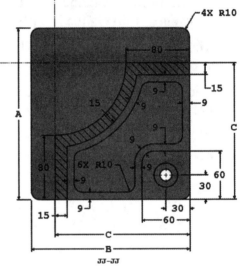

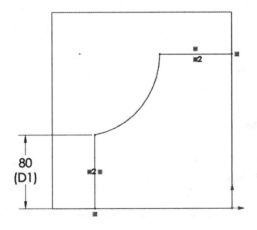

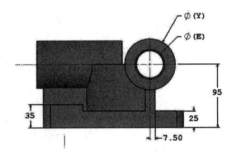

20. **Insert** the horizontal Global Variable C dimension as illustrated.

21. **Enter** C for Dimension Text.

22. **Insert** the vertical Global Variable C dimension.

23. **Enter** C for Dimension Text.

24. **Insert** a horizontal relation between the centerpoint of the arc and the horizontal sketch line end point if needed. The sketch is fully defined and is displayed in black.

25. **Insert** the Extrude-Thin feature. The Extrude-Thin feature is controlled by the Global Variable C and two dimensions that are equal (80mm). The overall height of the part is 95mm. Start from the Top face. Subtract 25mm from 95mm to obtain the correct depth for the Thin-Extrude feature. Click No in the Close Sketch With Model Edges dialog box. You want an open profile. If needed click the Reverse Direction in the Thin Feature box. Enter 15mm for Thickness. Enter 70mm for Depth in the Direction 1 box. Blind is the default End Conditions. Click OK from the Boss-Extrude PropertyManager. Extrude-Thin1 is displayed in the FeatureManager.

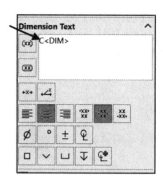

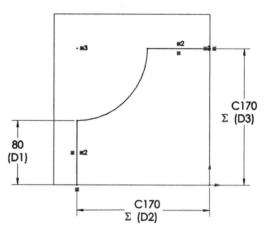

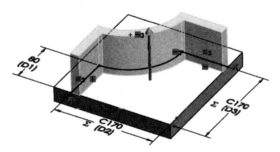

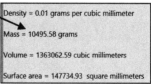

At this time, your model should have a mass of **10495.58 grams**. You should have the exact answer (within 1% of the stated value in the multiple choice section) before you move on to the next question.

Create the first cylinder (remember the two cylinders are not the same). The first cylinder outside diameter is controlled by equation X. X = A/3. The inside diameter is E = (41mm) and the depth is D = (130mm).

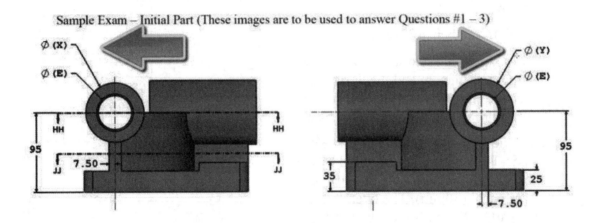

The first cylinder is offset 10mm from the Front Plane or face.

26. **Create** a Plane offset from the Front Plane (front face) 10mm. Plane1 is created.

27. **Create** Sketch3 on Plane1. Sketch a circle with the centerpoint Coincident at the midpoint of the Extrude-Thin1 feature. The dimension is driven by the X equation.

28. **Enter** the equation for X in the Modify dialog box.

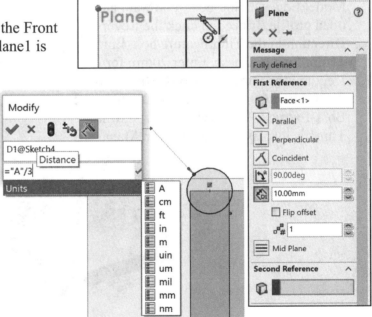

29. **Create** the Extruded Boss feature from Sketch3. Enter the extruded distance of the Global Variable D. Depth D = 130mm. Click Reverse direction if needed.

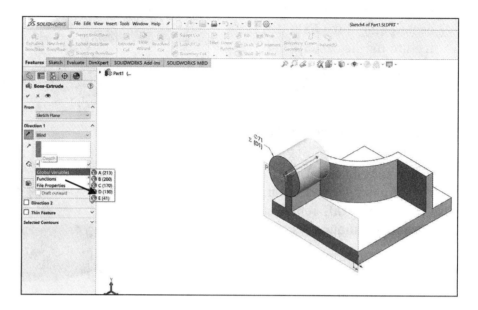

30. **Rebuild** the part. **View** the results in the Graphics window.

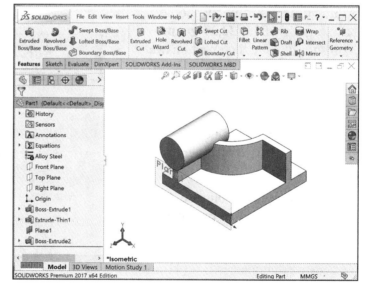

Prior to SOLIDWORKS 2013, a method commonly used to define variables for CSWP exam problems was to use Linked Values. Linked Values, also called linked dimensions, connect two or more dimensions without using equations or relations.

There are numerous ways to build the model in this section. A goal is to display different design intents and techniques.

31. **Create** Sketch4 on the front face of the cylinder to create the Extruded Cut feature. Use the Circle Sketch tool. Click the centerpoint of the Extruded Boss feature for the centerpoint.

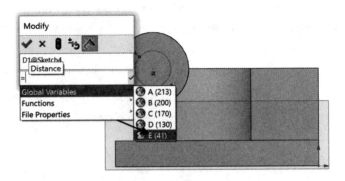

32. **Dimension** Sketch4. Insert the Global Variable E. E = 41mm.

33. **Enter** E for Dimension Text.

 There are numerous ways to build the model in this section. A goal is to display different design intents and techniques.

34. **Create** an Extruded Cut feature using Sketch4. Select Through All for End Condition to address any future design change for depth.

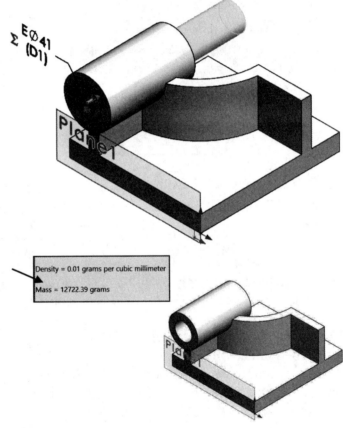

Density = 0.01 grams per cubic millimeter

Mass = 12722.39 grams

At this time, your model should have a mass of **12722.39 grams**. You should have the exact answer (within 1% of the stated value in the multiple choice section) before you move on to the next question.

Create the second cylinder (remember the two cylinders are _not_ the same).

The second cylinder outside diameter is controlled by equation Y.

Y = B/3 +10mm.

The inside diameter is E = (41mm).

The depth is D = (130mm).

The second cylinder is offset 10mm from either the Right Plane or right face.

35. **Create** a Plane (Plane2) offset from the Right Plane (right face) 10mm.

36. **Create** Sketch5 on Plane2. Sketch a circle with the centerpoint Coincident at the midpoint of the Extrude-Thin1 feature. The dimension is driven by the Y equation.

37. **Dimension** Sketch5. Enter the equation for Y in the Modify dialog box as displayed.

38. **Create** the Extruded Boss feature from Sketch5. Enter the extruded distance of the Global Variable D. Depth D = 137mm. Click Reverse direction if needed.

39. **Rebuild** the part. View the results in the Graphics window.

40. **Save** the part.

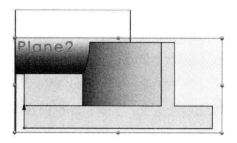

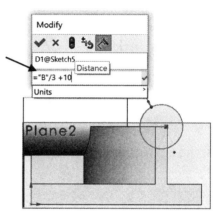

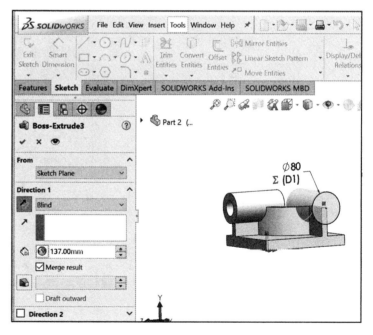

41. Create the Extruded Cut. **Create** Sketch6 on the Right face of the second cylinder. Use the Circle Sketch tool. Click the centerpoint of the Extruded Boss feature for the centerpoint.

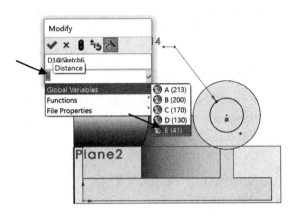

42. **Insert** the Global Variable E. E = 41mm.

43. **Enter** E for Dimension Text.

44. **Create** an Extruded Cut feature from Sketch6. Select Through All for End Condition to address any future design change in depth.

At this time, your model should have a mass of **15562.83 grams**.

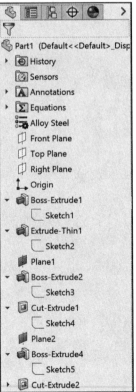

🔆 Always enter the needed decimal places in the answer field.

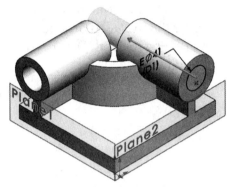

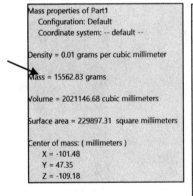

Create the Extruded Boss, Counterbore, and Fillet features on the right side of the model.

45. **Create** Sketch7 on the top face of Boss-Extrude1. Apply the Corner Rectangle tool from the right hand side. Insert an Equal geometric relation. Enter 60mm for dimension.

46. **Create** Boss-Extrude4 from Sketch7. Depth = 10mm. (35mm - 25mm) = 10mm from the provided information in the question.

Create the Counterbore hole using the Hole Wizard feature tool. Remember the provided information.

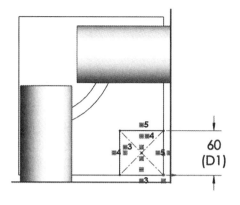

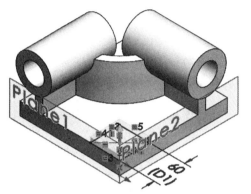

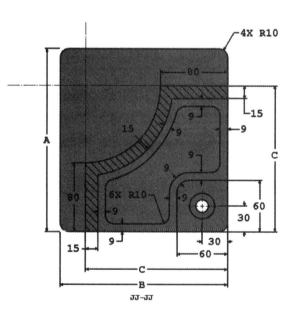

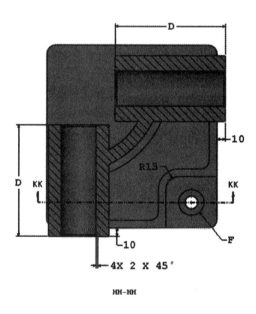

47. **Create** the Counterbore hole using the Hole Wizard on the top face of Boss-Extrude4.

There are numerous ways to build the models in this chapter. A goal is to display different design intents and techniques.

48. **Enter** the provided information in the Hole Specification PropertyManager. This is a Through All End Condition hole.

49. **Create** Sketch9. Use the Centerline Sketch tool with a Midpoint relation to location the center of the hole on the Boss-Extrude4 face. This saves time from creating two dimensions. The hole is complete.

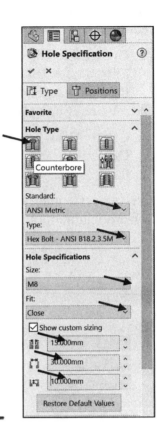

50. **Create** the first Constant radius Fillet (15mm) on the inside corner of the Boss-Extrude4 feature. You can use the Multiple radius fillet option to create all needed fillets, but for design intent and future modifications in the exam, insert two separate fillet features.

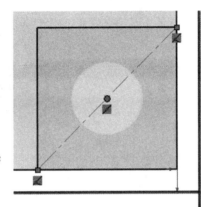

Use caution when taking the CSWP exam. Fillets are modified often. The Fillet PropertyManager was modified in SW 2016.

51. **Create** the second Constant radius Fillet (10mm) on the four outside edges of Boss-Extrude1 as illustrated.

At this time, your model should have a mass of **15729.69 grams**.

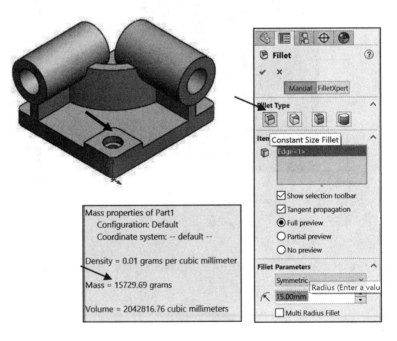

Mass properties of Part1
    Configuration: Default
    Coordinate system: -- default --

Density = 0.01 grams per cubic millimeter

Mass = 15729.69 grams

Volume = 2042816.76 cubic millimeters

Create the Extruded Cut feature (pocket) on the top face of Boss-Extrude1 using the Offset Entities Sketch tool; then create the Chamfer feature.

52. **Create** Sketch10. Use the Offset Entities Sketch (9mm Offset distance) tool. Note the direction of the offset.

53. **Create** the Cut-Extrude3 feature based on the bottom face of the model. Utilize the Offset from Surface End Condition. Enter 5mm from the provided model information.

At this time, your model should have a mass of **14198.79** grams.

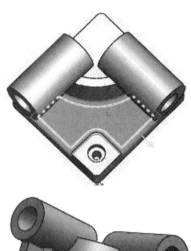

If you use faces vs. dimensions when creating features, this can help when you need to create a design change to maintain the design intent of the part.

54. **Insert** six (6) Constant radius Fillet features (10mm) on the Cut-Extrude3 feature per the provided information.

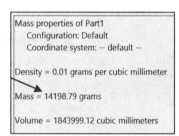

Mass properties of Part1
    Configuration: Default
    Coordinate system: -- default --

Density = 0.01 grams per cubic millimeter

Mass = 14198.79 grams

Volume = 1843999.12 cubic millimeters

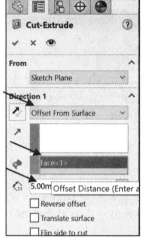

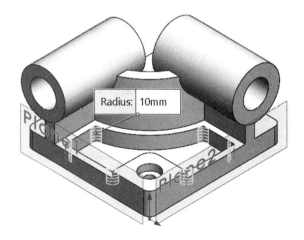

Radius: 10mm

Insert the last feature for the part, Chamfer1.

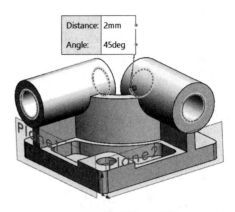

55. **Create** a Chamfer (Angle Distance) feature on the internal diameter edges of the cylinders. Angle = 45. Distance = 2mm. Select four edges. The last 2 edges are on the back of the cylinder.

56. **Calculate** the Mass Properties.

57. **Select 14207.34** grams for the answer in this section. The number matches the answer of d. You should be within 1% of the stated value before you move to the next section to modify the original part.

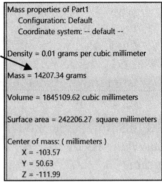

58. **Save** the part.

59. **Save as a copy Part2** for the second question. You are finished with this section.

Always enter the needed decimal places in the answer field.

This section presents a representation of the types of questions that you will see in this segment of the exam.

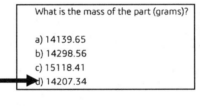

There might be slight variations in mass. These variations are still within the 1% required for the CSWP exam segments. Always save your models to verify the results.

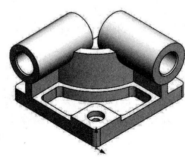

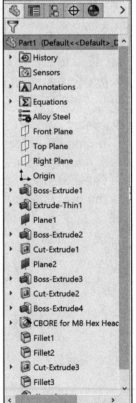

Segment 1 of the CSWP CORE exam - Second question

In this section, modify the original part using Global Variables. The material and units are the same but the variables A through E are different from their original values.

The Hole Wizard feature remains the same and the equation X and Y remain the same from the original part.

Read the question carefully. This section provides a single fill in the blank format, not a multiple choice format.

Provided Information:

Update parameters of the initial part.

Unit system: MMGS (millimeter, gram, second)

Decimal places: 2

Part origin: Arbitrary

Material: Alloy Steel

Density: 0.0077 g/mm^3

All holes through all unless shown otherwise.

Use the following parameters and equations which correspond to the dimensions labeled in the images:

A = 225 mm

B = 210 mm

C = 176 mm

D = 137 mm

E = 39 mm

F – Hole Wizard Standard: ANSI Metric - Counterbore

    Type: Hex Bolt - ANSI B18.2.3.5M

    Size: M8

    Fit: Close

    Through Hole Diameter: 15.00 mm

    Counterbore Diameter: 30.00 mm

    Counterbore Depth: 10.00 mm

    End Condition: Through All

CSWP R4 Segment 1

Question 2 of 6

For 25 points:

- Initial part

Build this part in SolidWorks. (Save part after each question in a different file in case it must be reviewed)
Unit system: MMGS (millimeter, gram, second)
Decimal places: 2
Part origin: Arbitrary
Material: 1060 Alloy Aluminum
Density = 0.0027 g/mm^3
All holes through all unless shown otherwise

-Use the following parameters and equations which correspond to the dimensions labeled in the images:

A = 135 mm
B = 58 mm
C = 180 mm
D = 26 mm
Y = 1.5 * B + 10mm

Hint #1: The dimensions that are to be linked or updated and are variable will be labeled with letters. Any dimensions that are simple value changes from one stage to another will be circled in the images.

Actual CSWP exam format

---

2. Update parameters of the initial part.

Unit system: MMGS (millimeter, gram, second)
Decimal places: 2
Part origin: Arbitrary
Material: Alloy Steel
Density = 0.0077 g/mm^3
All holes through all unless shown otherwise

Use the following parameters and equations which correspond to the dimensions labeled in the images:

A = 225 mm
B = 210 mm
C = 176 mm
D = 137 mm
E = 39 mm
F = Hole Wizard Standard: Ansi Metric Counterbore
    Type: Hex Bolt – ANSI B18.2.3.5M
    Size: M8
    Fit: Close
    Through Hole Diameter: 15.00 mm
    Counterbore Diameter: 30.00 mm
    Counterbore Depth: 10.00 mm
    End Condition: Through All
X = A/3
Y = B/3 + 10mm

X = A/3

Y = B/3 + 10mm

Hint #1: The dimensions that are to be linked or updated and are variable will be labeled with letters. Any dimensions that are simple value changes from one stage to another will be circled in the images.

Hint #2: To save the most time, make use of linked dimensional values and equations.

Measure the mass of the part.

What is the mass of the part (grams)?

SOLIDWORKS displays a circle, ellipse or a square around the areas and features that require modification from the original part.

This section presents a representation of the types of questions that you will see in this segment of the exam.

The images displayed on the exam are not to scale due to differences in the parts being downloaded for each tester.

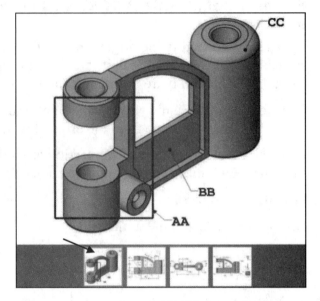

Actual CSWP exam picture

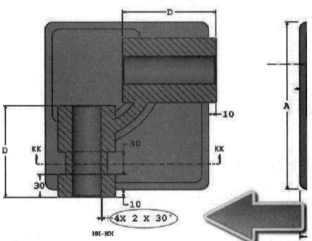

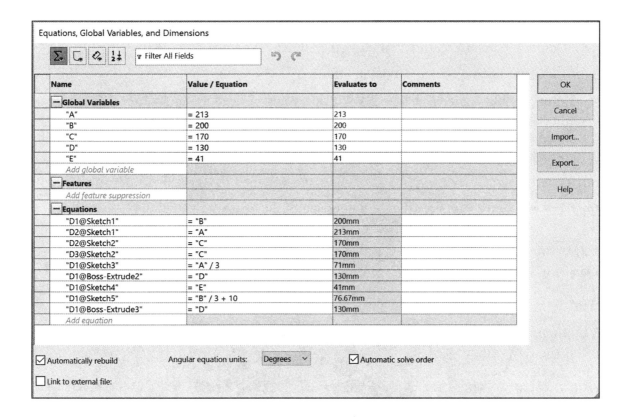

SOLIDWORKS 2017 screen shot Part 1

Let's begin.

Modify the variables that are different from Part1.

1. **Display** the Equations, Global Variables, and Dimensions dialog box.

2. **Enter** the five new Global Variables (A, B, C, D & E) as illustrated.

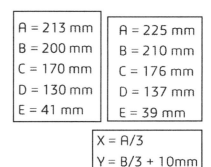

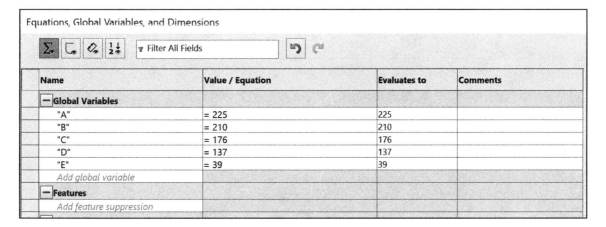

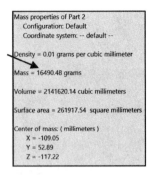

Double-click on a feature to modify the Global Variable.

3. **Calculate** the mass of the model in grams.

4. **Enter 16490.48** grams. In the CSWP exam you will need to enter this number exactly. You need to be within 1% of the stated value in the single answer format to get this question correct.

5. **Save** the part.

6. **Save as a copy Part3** for the third question of the exam.

Always enter the needed decimal places (in this case 2) in the answer field.

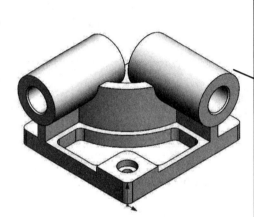

You should have the exact answer (within 1% of the stated value in the multiple choice section) before you move on to the next question.

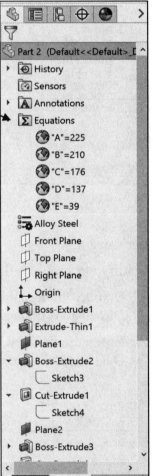

Segment 1 of the CSWP CORE exam - Third question

Update various parameters of the part. Modify the dimensions under Global Variables. Read the question carefully. Identify what variables and equations are the same vs. different. Has the material changed? Did a feature change? A question will be presented to you in multiple steps and you need to get each step correct to get the question correct. Provided Information:

Update parameters of the initial part.

Unit system: MMGS (millimeter, gram, second)

Decimal places: 2

Part origin: Arbitrary

Material: Alloy Steel

Density: 0.0077 g/mm^3

All holes through all unless shown otherwise.

Use the following parameters and equations which correspond to the dimensions labeled in the images:

A = 209 mm

B = 218 mm

C = 169 mm

D = 125 mm

E = 41 mm

F = Hole Wizard Standard: ANSI Metric - Counterbore

    Type: Hex Bolt - ANSI B18.2.3.5M

    Size: M8

    Fit: Close

    Through Hole Diameter: 15.00 mm

    Counterbore Diameter: 30.00 mm

    Counterbore Depth: 10.00 mm

    End Condition: Through All

X = A/3

Y = B/3 + 10mm

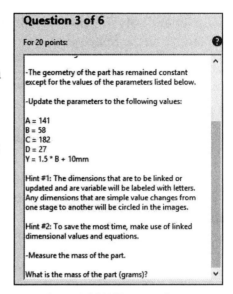

Question 3 of 6

For 20 points:

-The geometry of the part has remained constant except for the values of the parameters listed below.

-Update the parameters to the following values:

A = 141
B = 58
C = 182
D = 27
Y = 1.5 * B + 10mm

Hint #1: The dimensions that are to be linked or updated and are variable will be labeled with letters. Any dimensions that are simple value changes from one stage to another will be circled in the images.

Hint #2: To save the most time, make use of linked dimensional values and equations.

-Measure the mass of the part.

What is the mass of the part (grams)?

Actual CSWP exam format

3. Update parameters of the initial part.

Unit system: MMGS (millimeter, gram, second)
Decimal places: 2
Part origin: Arbitrary
Material: Alloy Steel
Density = 0.0077 g/mm^3
All holes through all unless shown otherwise

Use the following parameters and equations which correspond to the dimensions labeled in the images:

A = 209 mm
B = 218 mm
C = 169 mm
D = 125 mm
E = 41 mm
F = Hole Wizard Standard: Ansi Metric Counterbore
    Type: Hex Bolt – ANSI B18.2.3.5M
    Size: M8
    Fit: Close
    Through Hole Diameter: 15.00 mm
    Counterbore Diameter: 30.00 mm
    Counterbore Depth: 10.00 mm
    End Condition: Through All
X = A/3
Y = B/3 + 10mm

Hint #1: The dimensions that are to be linked or updated and are variable will be labeled with letters. Any dimensions that are simple value changes from one stage to another will be circled in the images.

Hint #2: To save the most time, make use of linked dimensional values and equations.

Measure the mass of the part.

What is the mass of the part (grams)?

To display all dimensions, right-click the Annotations folder from the FeatureManager and check the Display Annotations box.

Let's begin.

1. **Display** the Equations, Global Variables, and Dimension dialog box.

2. **Enter** the five Global Variables (A, B, C, D & E) as illustrated. Modify the variables that are different.

A = 225 mm	A = 209 mm
B = 210 mm	B = 218 mm
C = 176 mm	C = 169 mm
D = 137 mm	D = 125 mm
E = 39 mm	E = 41 mm

| X = A/3 |
| Y = B/3 + 10mm |

The equations ($X = A/3$, $Y = B/3 + 10mm$) have not been modified between the first, second or third question.

The CSWP exam in this section provides variables that either increase or decrease from the original part question. Design for this during the exam.

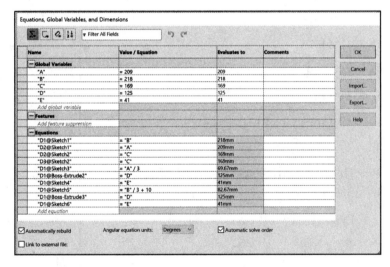

3. **Calculate** the mass of the model in grams.

4. **Enter 15100.47** grams.

5. **Save** the part.

6. **Rename** Part3 to Part4 for the fourth question of the exam.

Enter the needed decimal places (in this case 2) in the answer field.

You should have the exact answer (within 1% of the stated value in the multiple choice section) before you move on to the next question.

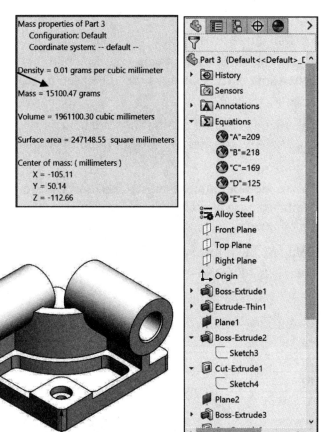

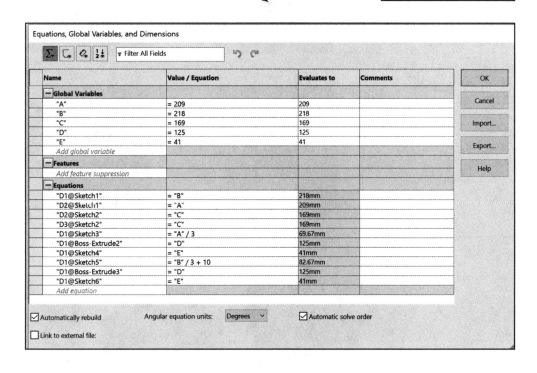

Segment 1 of the CSWP CORE exam - Fourth question

Stage 2: Modify the part using the following dimensions. (These images are to be used to answer questions 4 and 5.)

The changes from the initial part are concentrated in areas AA, BB and CC shown in the first two images.

The needed modifications AA, BB, and CC are displayed with a circle, ellipse or a square in the CSWP exam.

Compare the information with your existing part. This question provides a multiple choice answer.

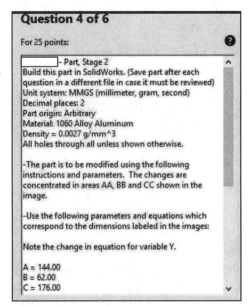

Question 4 of 6

For 25 points:

[ ] - Part, Stage 2
Build this part in SolidWorks. (Save part after each question in a different file in case it must be reviewed)
Unit system: MMGS (millimeter, gram, second)
Decimal places: 2
Part origin: Arbitrary
Material: 1060 Alloy Aluminum
Density = 0.0027 g/mm^3
All holes through all unless shown otherwise.

-The part is to be modified using the following instructions and parameters. The changes are concentrated in areas AA, BB and CC shown in the image.

-Use the following parameters and equations which correspond to the dimensions labeled in the images:

Note the change in equation for variable Y.

A = 144.00
B = 62.00
C = 176.00

Actual CSWP exam format

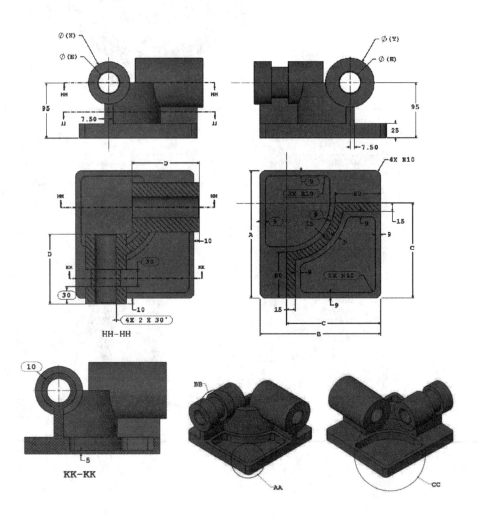

Provided Information:

4. Stage 2

Unit system: MMGS (millimeter, gram, second)

Decimal places: 2

Part origin: Arbitrary

Material: Alloy Steel

Density: 0.0077 g/mm^3

All holes through all unless shown otherwise.

Use the following parameters and equations which correspond to the dimensions labeled in the images:

A = 221 mm

B = 211 mm

C = 165 mm

D = 121 mm

E = 37 mm

X = A/3

Y = B/3 + 15mm

Note: The equation for Y has changed from the initial part.

Hint #1: The dimensions that are to be linked or updated and are variable will be labeled with letters. Any dimensions that are simple value changes from one stage to another will be circled in the images.

Hint #2: To save the most time, make use of linked dimensional values and equations.

Measure the mass of the part.

What is the mass of the part (grams)?

a) 13095.40

b) 13206.40

c) 13313.35

d) 13395.79

---

4. Stage 2 – Modify

Modify the part using the following dimensions.

Note: The changes from the initial part are concentrated in areas AA, BB and CC shown in the images.

Unit system: MMGS (millimeter, gram, second)
Decimal places: 2
Part origin: Arbitrary
Material: Alloy Steel
Density = 0.0077 g/mm^3
All holes through all unless shown otherwise

Use the following parameters and equations which correspond to the dimensions labeled in the images:

A = 221 mm
B = 211 mm
C = 165 mm
D = 121 mm
E = 37 mm
X = A/3
Y = B/3 + 15mm

Note: The equation for Y has changed from the initial part.

The material and units are the same. Equation A is the same. Variables A through E are different. The Hole Wizard feature is removed with a few other features as illustrated. Equation Y is different. Read the question carefully.

Address AA modification in the model. Create a single pocket and remove the Hole Wizard feature along with the Boss-Extrude4 feature and a few fillets. Recover and repair from missing items.

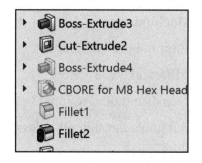

Let's begin.

1. **Suppress** Fillet1, CBORE and Boss-Extrude4. A dialog box is displayed.

2. **Press** the Stop and Repair button from the SOLIDWORKS dialog box. Fillet3 has a reference issue. There are missing items in the existing feature.

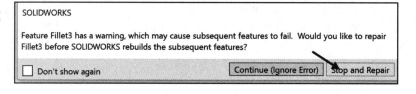

As a general rule, suppress features before you delete them. This will inform you if there are any rebuild or feature errors during modification in the exam.

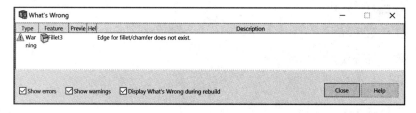

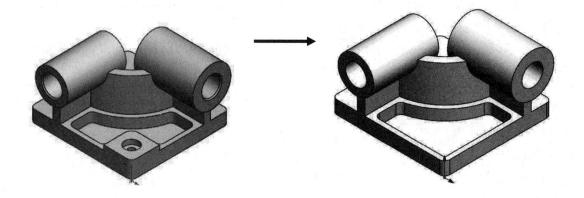

3. **Edit** the Fillet3 feature and repair. Delete any missing edges and insert the new edge.

But wait, you can't insert the new edge (as illustrated) because it is part of the offset from the original sketch when you created the part.

Create a square corner for Sketch10 to select the needed edge for Fillet3.

4. **Edit** Sketch10 from the Cut-Extrude3 feature. Delete the fillet.

5. **Insert** the needed edge for Fillet3 using the Trim Entities sketch tool (Corner option).

6. **Edit** the Fillet3 feature and add the needed edge that you just created. Verify that you have 5 fillets with the radius of 10mm. There are no errors displayed in the FeatureManager.

7. **Delete** all suppressed features and any unneeded sketch in the FeatureManager: Fillet1, CBORE, Boss-Extrude4 and Sketch7. Suppress features before you delete them. This will inform you if there are any rebuild or feature errors during modification in the exam.

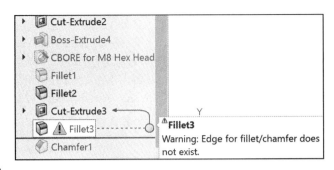

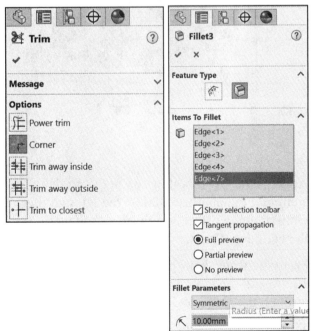

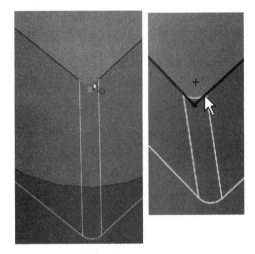

8. **Roll** back the Rollback bar in the FeatureManager.

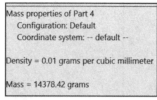

9. **Calculate** the mass of the model in grams.

10. Your mass at this time should be **14378.42 grams**.

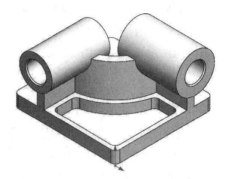

You should have the exact answer (within 1% of the stated value in the multiple choice section) before you move on to the next question.

11. **Save** the part.

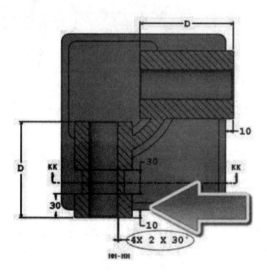

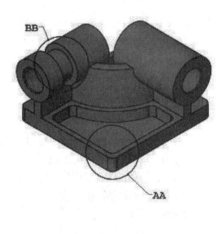

Address the BB modification in the model. Create a circular cut as illustrated on the left cylinder. The circular cut is offset 30mm from the front face of the cylinder. The depth of the cut is 30mm.

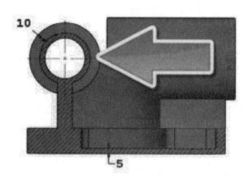

From the front view, the circular cut is offset 10mm and the circular cut does not go completely through the Extrude-Thin1 feature.

Then modify the Chamfer feature of the cylinders.

In this section utilize the Offset Entities, Convert Entities, Line and Trim Sketch tools to create the Base Sketch for the Extruded Cut feature.

Let's begin.

12. **Create** a Sketch plane (30mm) offset from the front face of the cylinder as illustrated. Plane3 is displayed in the FeatureManager.

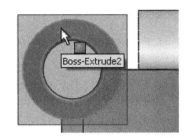
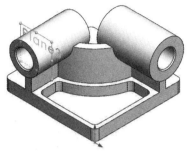

13. **Create** the Base Sketch on Plane3. Apply the Convert Entities Sketch tool to utilize the outside cylindrical geometry of the tube.

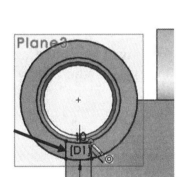

14. **Apply** the Offset Entities Sketch tool with an offset distance of 10mm. Click the outside cylindrical edge and reverse the direction if needed. You created an inside and outside ring for the Extruded Cut feature on your Base Sketch (Sketch11).

15. **Create** two vertical lines from the outside cylindrical edge to the inside cylindrical edge of the ring. The sketch is fully defined.

16. **Trim** any unwanted sketch geometry to finish the sketch for the Extruded Cut feature. Your Base sketch should consist of two arcs and two vertical lines.

17. **Create** an Extruded Cut feature with a depth of 30mm. The Extruded Cut feature removes the needed material and keeps the Extrude-Thin1 feature unbroken.

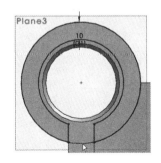

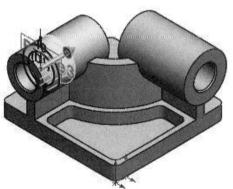

Next address the modification of the Chamfer feature in the front face of the cylinder.

18. **Modify** the cylinder Chamfer feature from 45 degrees to 30 degrees.

At this time your model should have a mass of **13983.95 grams**.

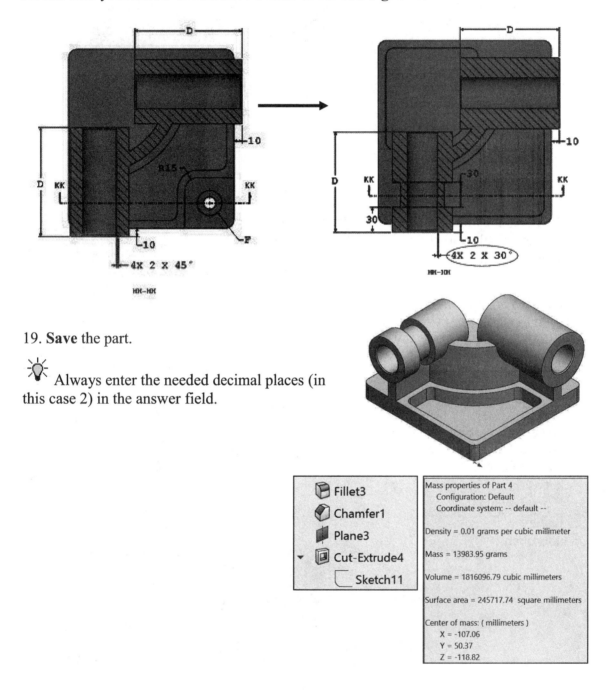

19. **Save** the part.

Always enter the needed decimal places (in this case 2) in the answer field.

Mass properties of Part 4
    Configuration: Default
    Coordinate system: -- default --

Density = 0.01 grams per cubic millimeter

Mass = 13983.95 grams

Volume = 1816096.79 cubic millimeters

Surface area = 245717.74 square millimeters

Center of mass: (millimeters)
    X = -107.06
    Y = 50.37
    Z = -118.82

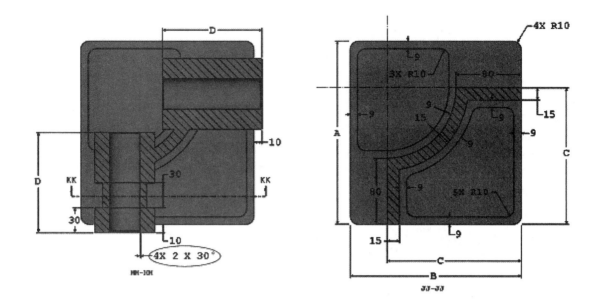

Address the Extruded Cut and fillet feature and then address the Global Variables A through E and equation Y.

There are numerous ways to create the sketch for the Extruded Cut feature. In this case, utilize Construction geometry. Construction geometry helps you create a sketch but is not part of the feature.

20. **Create** a sketch using the Offset Entities Sketch tool on the top back face. Enter 9mm for Offset distance. Reverse the direction if needed.

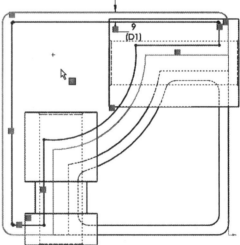

21. **Window-select** the part. Check the For construction box. Again, construction geometry helps create a sketch but is not part of the feature.

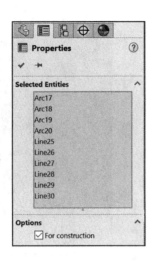

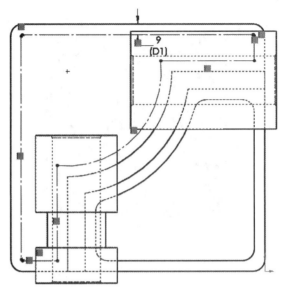

Create a 90^0 Arc and inference the vertical line from the Extrude-Thin1 feature.

22. **Create** a 90^0 Arc. Use the Centerpoint Arc Sketch tool. Click the centerpoint as illustrated, and drag directly to the right until you see the inference symbol on the construction arc. Click the start point. Drag downward and click the end point to create the 90^0 Arc. The mouse pointer displays A = 90 for angle feedback.

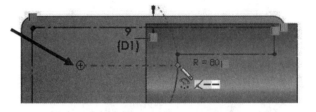

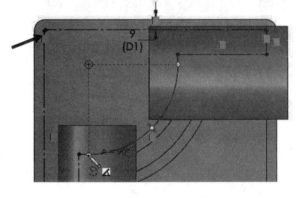

Sometimes when you convert sketch geometry, unwanted arcs and points are created. In the next section, delete any unwanted sketch geometry to cleanly create Sketch12.

23. **Remove** the top left fillet with the Trim Entities Sketch tool.

24. **Restore** the corner of the fillet feature with the Trim Entities Sketch tool (Corner option) or by just dragging the endpoints together.

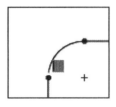

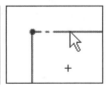

25. **Remove** all unwanted sketch geometry around the end point of the 90^0 Arc.

Now you can utilize the horizontal and vertical construction geometry with the 9mm offset with new lines and the Arc is created correctly. Again, there are other ways to create the sketch for the Extruded Cut feature.

26. **Complete** the close sketch profile with the Line Sketch tool.

Insert all needed Geometric relations.

27. **Insert** a vertical relation between the end point of the Arc and the corner point of the Thin-Extrude1 feature. The sketch should be fully defined and displayed in black.

28. **Create** the Extruded Cut feature. Apply the Up To Surface End Condition and click the inside face of Cut-Extrude3. The two surfaces provide a similar dimension.

29. **Apply** the Fillet feature. Insert three fillets per the provided drawing. 10mm is the fillet radius.

At this time your model should have a mass of **12154.09 grams**.

Mass = 12154.09 grams

Volume = 1578453.44 cubic millimeters

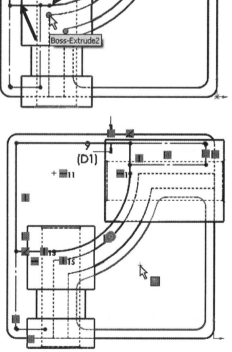

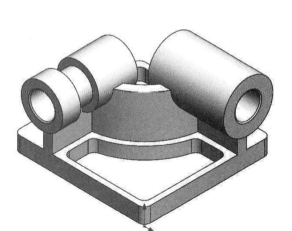

30. **Display** the Equations, Global Variables and Dimension dialog box.

31. **Enter** the five new Global Variables (A, B, C, D & E) and the new Y equation as illustrated.

> A = 221 mm
> B = 211 mm
> C = 165 nm
> D = 121 mm
> E = 37 mm
> X = A/3
> Y = B/3 + 15mm
>
> Note: The equation for Y has changed from the initial part.

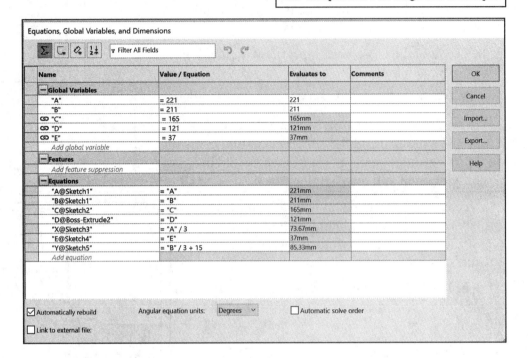

32. **Calculate** the mass of the model.

33. **Select 13206.40 grams**. The number matches the answer of b. You should be within 1% of the stated value before you move to the next section to modify the part.

34. **Save** the part.

35. **Rename** Part4 to Part5.

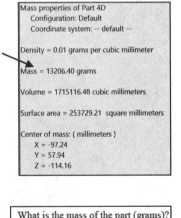

Mass properties of Part 4D
   Configuration: Default
   Coordinate system: -- default --

Density = 0.01 grams per cubic millimeter

Mass = 13206.40 grams

Volume = 1715116.48 cubic millimeters

Surface area = 253729.21 square millimeters

Center of mass: (millimeters)
   X = -97.24
   Y = 57.94
   Z = -114.16

🔅 Always save your models to verify your results.

🔅 Use the Comments box to label your equations.

🔅 Confirm that your math is correct. Use the Measure tool during the exam.

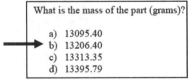

What is the mass of the part (grams)?

  a) 13095.40
  b) 13206.40
  c) 13313.35
  d) 13395.79

Segment 1 of the CSWP CORE exam - Fifth question

Compare the provided information with your existing part. The Global Variables A through E change and the equations are the same from the last question but equation Y has changed from the initial part. This question provides a fill in the blank format.

Provided Information:

5. Stage 2 - Update Parameters

Unit system: MMGS (millimeter, gram, second)

Decimal places: 2

Part origin: Arbitrary

Material: Alloy Steel

Density: 0.0077 g/mm^3

All holes through all unless shown otherwise.

Use the following parameters and equations which correspond to the dimensions labeled in the images:

A = 229 mm

B = 217 mm

C = 163 mm

D = 119 mm

E = 34 mm

X = A/3

Y = B/3 + 15mm

Hint #1: The dimensions that are to be linked or updated and are variable will be labeled with letters. Any dimensions that are simple value changes from one stage to another will be circled in the images.

Hint #2: To save the most time, make use of linked dimensional values and equations.

Measure the mass of the part.

What is the mass of the part (grams)?

Let's begin.

1. **Display** the Equations, Global Variables and Dimensions dialog box.

2. **Enter** the five new Global Variables (A, B, C, D & E) as illustrated.

A = 229 mm	
B = 217 mm	
C = 163 mm	
D = 119 mm	
E = 34 mm	
X = A/3	
Y = B/3 + 15mm	

Equations, Global Variables, and Dimensions

Filter All Fields

Name	Value / Equation	Evaluates to	Comments	
Global Variables				OK
"A"	= 229	229		
"B"	= 217	217		Cancel
"C"	= 163	163		
"D"	= 119	119		Import...
"E"	= 34	34		
Add global variable				Export...
Features				
Add feature suppression				Help
Equations				
"D1@Sketch1"	= "B"	217mm		
"D2@Sketch1"	= "A"	229mm		
"D2@Sketch2"	= "C"	163mm		
"D3@Sketch2"	= "C"	163mm		
"D1@Sketch3"	= "A" / 3	76.33mm	Equation X	
"D1@Boss-Extrude2"	= "D"	119mm		
"D1@Sketch4"	= "E"	34mm		
"D1@Sketch5"	= "B" / 3 + 15	87.33mm	Equation Y	
"D1@Boss-Extrude3"	= "D"	119mm		
"D1@Sketch6"	= "E"	34mm		
Add equation				

☐ Automatically rebuild   Angular equation units: Degrees ▼   ☑ Automatic solve order

☐ Link to external file:

3. **Calculate** the mass of the part.

4. **Enter 14208.01 grams**.

5. **Save** the part.

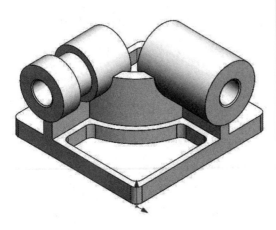

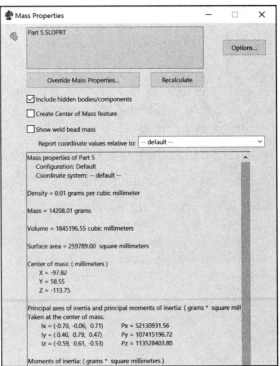

Mass Properties

Part 5.SLDPRT

Options...

Override Mass Properties...   Recalculate

☑ Include hidden bodies/components

☐ Create Center of Mass feature

☐ Show weld bead mass

Report coordinate values relative to: -- default --

Mass properties of Part 5
  Configuration: Default
  Coordinate system: -- default --

Density = 0.01 grams per cubic millimeter

Mass = 14208.01 grams

Volume = 1845196.55 cubic millimeters

Surface area = 259789.00 square millimeters

Center of mass: (millimeters)
  X = -97.82
  Y = 58.55
  Z = -113.75

Principal axes of inertia and principal moments of inertia: (grams * square mill
Taken at the center of mass.
  Ix = (-0.70, -0.06, 0.71)   Px = 52130931.56
  Iy = (0.40, 0.79, 0.47)   Py = 107415196.72
  Iz = (-0.59, 0.61, -0.53)   Pz = 113528403.80

Moments of inertia: (grams * square millimeters)

Segment 1 of the CSWP CORE exam - Additional Practice Problems

In this section, there are fewer step-by-step procedures. Use the provided initial and final models with the rollback bar if needed.

Question 1:
Build the following part in SOLIDWORKS.

Provided information:

Units: MMGS (millimeter, gram, second)

Decimal Places: 2

Part Origin: Arbitrary

Material: 6061 Alloy

Density: 2700 kg/m^2

All holes through all unless shown otherwise.

Use the following parameters and equations which correspond to the dimensions labeled in the images:

A: 60 mm

B: 45 mm

C: 52 degrees

D: 20 mm

X: 175 + D/2

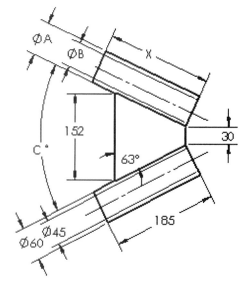

What is the mass of the part in (grams)?

a) 686.50

b) 1858.61

c) 1845.10

d) 1742.88

You should have the exact answer (within 1% of the stated value in the multiple choice section) before you move on to the next question.

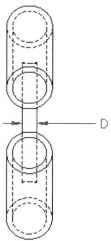

☀ There are numerous ways to build the model in this section. A goal is to display different design intents and techniques.

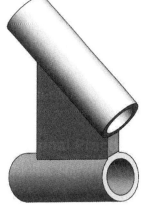

Let's begin.

Create a new part in SOLIDWORKS.

1. **Create** a folder to save your models.

2. **Create** a new part.

3. **Set** document properties (drafting standard, units and precision) for the model.

4. **Assign** 6061 Alloy material.

In this problem, first address Global Variables and Equations.

5. **Display** the Equations, Global Variables, and Dimensions dialog box.

6. **Enter** the below information for the provided parameters and equations as illustrated.

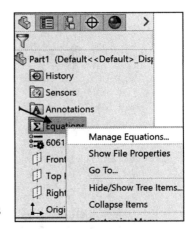

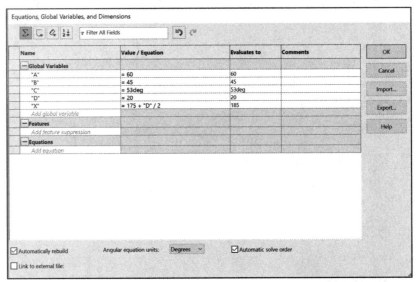

7. **Click** OK from the dialog box.

8. **Expand** the Equations folder in the FeatureManager. View the results.

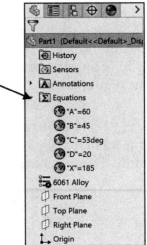

Start the Base feature (Boss-Extrude1) with Sketch1.

9. **Create** Sketch1. Select the Front plane and utilize Global Variables as illustrated.

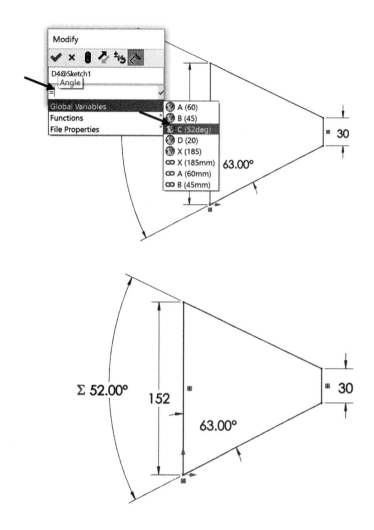

10. **Create** Boss-Extrude1. Apply the Mid-Plane End Condition with a distance of 20mm. Link the extruded distance to the Global Variable D. At this time your mass is 614.36 grams.

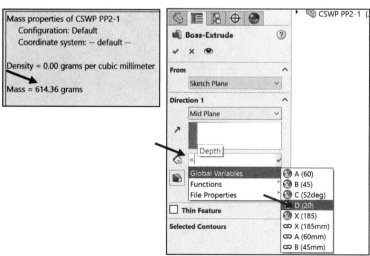

11. **Create** Sketch2 on the Front plane for the Revolve1 feature. Note the geometric relations in the sketch. There are small line segments in this sketch. Zoom in to select correct references.

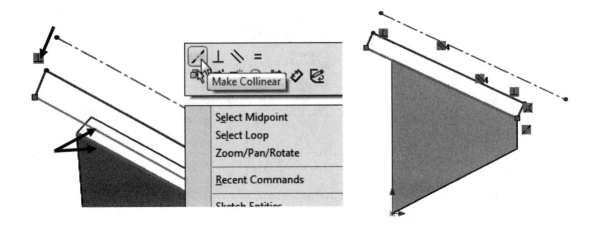

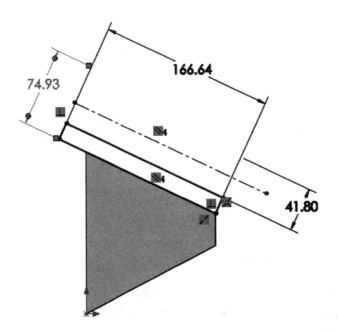

12. **Utilize** Global Variables. Note: There are many different ways to create this model.

   The CSWP exam in this section provides variables that either increase or decrease from the original part question. Design for this during the exam.

13. **Create** Revolve1. Revolve1 is the top tube feature as illustrated. At this time your model should have a mass of **1232.24 grams**.

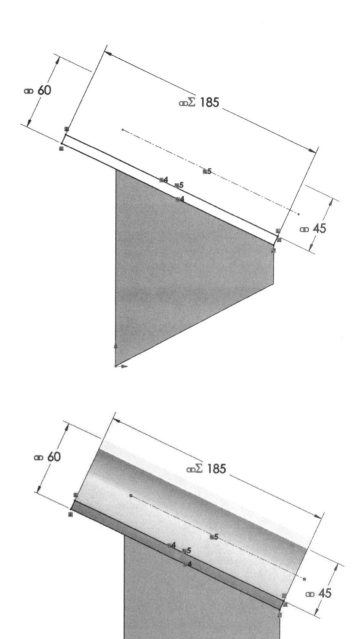

Merge the two bodies.

14. **Create** Sketch3. Select the top face of Boss-Extrude1. Utilize the Convert entities Sketch tool.

15. **Create** Boss-Extrude2. Extrude Up to Next (End Condition) making a single (Merge results) body part. At this time your model should have a mass of **1236.45 grams**.

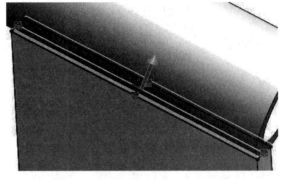

Always enter the needed decimal places (in this case 2) in the answer field.

Confirm that your math is correct. Use the Measure tool during the exam.

16. **Save** the part.

17. **Name** the part CSWP PP2-1.

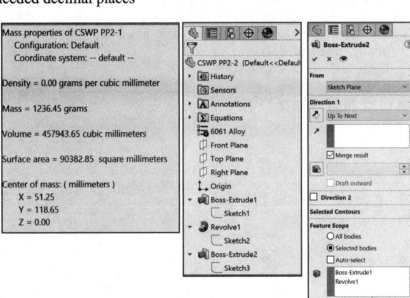

Mass properties of CSWP PP2-1
    Configuration: Default
    Coordinate system: -- default --

Density = 0.00 grams per cubic millimeter

Mass = 1236.45 grams

Volume = 457943.65 cubic millimeters

Surface area = 90382.85 square millimeters

Center of mass: (millimeters)
    X = 51.25
    Y = 118.65
    Z = 0.00

Create the bottom tube with a second
Revolve feature.

18. **Create** Sketch4 on the Front plane for
the Revolve2 feature. Note the
geometric relations in the sketch. There
are small line segments in this sketch.
Zoom in to select the correct references.
Inert all needed relations and
dimensions.

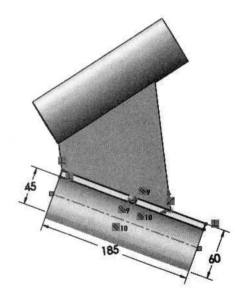

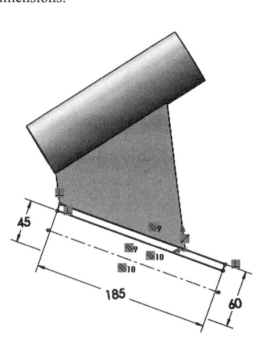

19. **Create** Revolve2. Revolve2 is the bottom tube
feature as illustrated. At this time your model
should have a mass of
1854.33 grams.

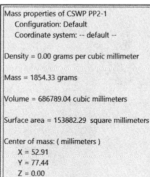

Mass properties of CSWP PP2-1
    Configuration: Default
    Coordinate system: -- default --

Density = 0.00 grams per cubic millimeter

Mass = 1854.33 grams

Volume = 686789.04 cubic millimeters

Surface area = 153882.29 square millimeters

Center of mass: (millimeters)
    X = 52.91
    Y = 77.44
    Z = 0.00

Merge the two bodies.

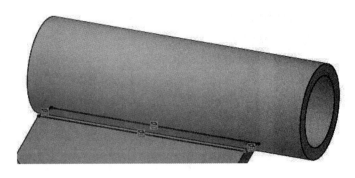

20. **Create** Sketch5. Select the bottom face of Boss-Extrude1. Utilize the Convert Entities Sketch tool.

21. **Create** Boss-Extrude3. Extrude Up to Next (End Condition) making a single (Merge results) body part.

22. **Calculate** the mass of the model in grams.

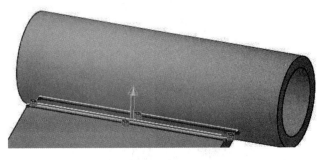

23. **Enter 1858.61** grams. In the CSWP exam you will need to enter this number exactly. You need to be within 1% of the stated value in the single answer format to get this question correct.

24. **Save** the part.

25. **Rename** the CSWP PP2-1 part to CSWP PP2-2.

What is the mass of the part in (grams)?
a) 686.50

b) 1858.61

c) 1845.10

d) 1742.88

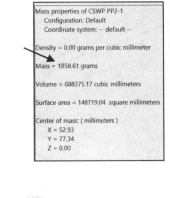

🔆 This section presents a representation of the types of questions that you will see in this segment of the exam.

🔆 At the time of this writing, there were slight variations in Mass Properties between SOLIDWORKS versions. These variations are still within the 1% required for the CSWP exam segments. Always save your models to verify your results.

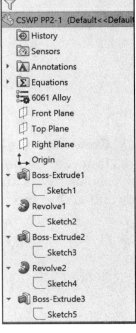

Question 1: Part II

Compare the provided information with the existing part (CSWP PP2-2) which you created in the previous section. Use the provided initial and final models with the rollback bar if needed.

What is the mass of the part in grams?

Provided information:

Units: MMGS (millimeter, gram, second)

Decimal Places: 2

Part Origin: Arbitrary

Material: 6061 Alloy

Density: 2700 kg/m^2

All holes through all unless shown otherwise.

Use the following parameters and equations which correspond to the dimensions labeled in the images.

A: 80 mm

B: 60 mm

C: 50 degree

D: 40 mm

X: 180 + D/2

You should have the exact answer (within 1% of the stated value in the multiple choice section) before you move on to the next question.

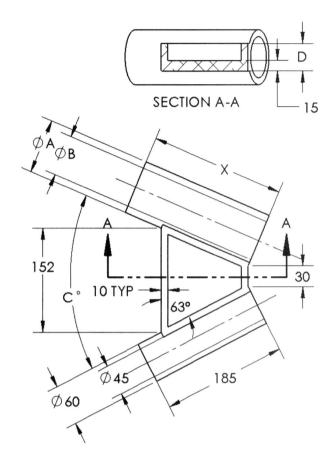

SECTION A-A

There are numerous ways to build the model in this section. A goal is to display different design intents and techniques.

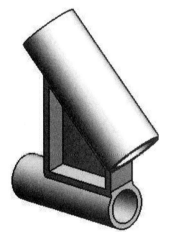

Let's begin.

Compare the provided **dimensions**, **materials**, **variables** and **equation**.

Question 1:
Build the following part in SOLIDWORKS.

Provided information:

Units: MMGS (millimeter, gram, second)

Decimal Places: 2

Part Origin: Arbitrary

Material: 6061 Alloy

Density: 2700 kg/m²

All holes through all unless shown otherwise.

Use the following parameters and equations which correspond to the dimensions labeled in the images:

A: 60 mm

B: 45 mm

C: 52 degrees

D: 20 mm

X: 175 + D/2

What is the mass of the part in (grams)?

a) 686.50

b) 1858.61

c) 1845.10

d) 1742.88

→

Question 1: Part II

Compare the provided information with the existing part (CSWP PP2-2) which you created in the previous section. Use the provided initial and final models with the rollback bar if needed.

What is the mass of the part in grams?

Provided information:

Units: MMGS (millimeter, gram, second)

Decimal Places: 2

Part Origin: Arbitrary

Material: 6061 Alloy

Density: 2700 kg/m²

All holes through all unless shown otherwise.

Use the following parameters and equations which correspond to the dimensions labeled in the images.

A: 80 mm

B: 60 mm

C: 50 degree

D: 40 mm

X: 180 + D/2

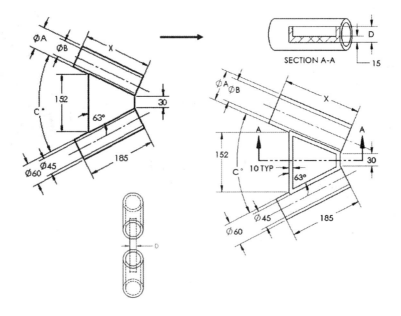

Where do you start? Three variables and the equation were changed. Edit the Equations, Global Variables, and Dimensions dialog box.

A: 60 mm	A: 80 mm
B: 45 mm	B: 60 mm
C: 52 degrees	C: 50 degree
D: 20 mm	D: 40 mm
X: 175 + D/2	X: 180 + D/2

1. **Right-click** the Equations folder. Click Manager Equations. The Equations, Global Variables, and Dimensions dialog box is displayed.

2. **Enter** the below information for the provided parameters and equation as illustrated.

Equations, Global Variables, and Dimensions

▼ Filter All Fields

Name	Value / Equation	Evaluates to	Comments	
Global Variables				OK
∞ "A"	= 80	80mm		
∞ "B"	= 60	60mm		Cancel
∞ "C"	= 50deg	50deg		
"D"	= 40	40		Import...
∞ "X"	= 180 + "D" / 2	200mm		
Add global variable				Export...
Features				
Add feature suppression				Help
Equations				
"D1@Boss-Extrude1"	= "D"	40mm		
Add equation				

☑ Automatically rebuild Angular equation units: Degrees ∨ ☑ Automatic solve order

☐ Link to external file:

---

3. **Click** OK from the dialog box.

At this time your model should have a mass of **3153.76 grams**.

4. **Expand** the Equations folder. View the results.

Address the Offset Extruded Cut feature displayed in the illustration.

5. **Create** Sketch6 on the front face of the plate. Utilize the Offset Entities Sketch tool. Enter 10mm depth to the inside.

6. **Display** an Isometric view to create the Extruded Cut feature.

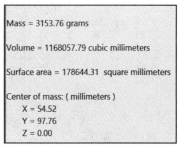

Mass = 3153.76 grams

Volume = 1168057.79 cubic millimeters

Surface area = 178644.31 square millimeters

Center of mass: (millimeters)
   X = 54.52
   Y = 97.76
   Z = 0.00

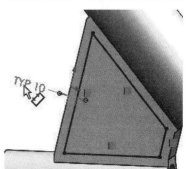

7. **Create** Cut-Extrude1. Utilize the Offset from Surface End Condition. Select the back face. Enter 15mm (Depth) as illustrated.

8. **Save** the part.

9. **Calculate** the mass of the model in grams.

What is the mass of the part in grams? Check your work.

10. **Enter 2531.70 grams**.

11. **Rename** part CSWP PP2-2 to CSWP PP2-3.

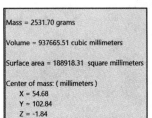

Mass = 2531.70 grams

Volume = 937665.51 cubic millimeters

Surface area = 188918.31 square millimeters

Center of mass: (millimeters)
  X = 54.68
  Y = 102.84
  Z = -1.84

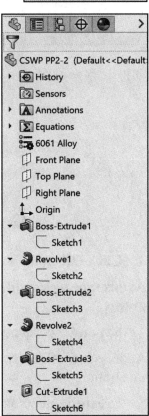

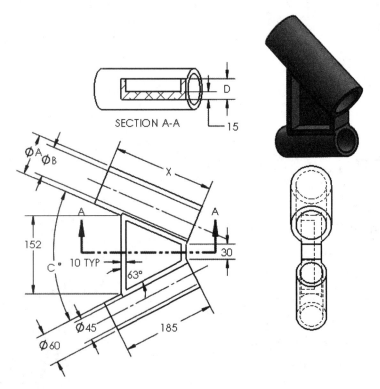

SECTION A-A

Question 1: Part III

Build the following part in SOLIDWORKS.

Compare the provided information (dimensions, material, variables and equation) with the existing part (CSWP PP2-3) which you created in the previous section.

What is the mass of the part in grams?

Provided Information:

Units: MMGS (millimeter, gram, second)

Decimal Places: 2

Part Origin: Arbitrary

Material: 6061 Alloy

Density: 2700 kg/m^2

All holes through all unless shown otherwise.

Use the following parameters and equations which correspond to the dimensions labeled in the images.

A: 80 mm

B: 60 mm

C: 50 degrees

D: 40 mm

X: 180 + D/2

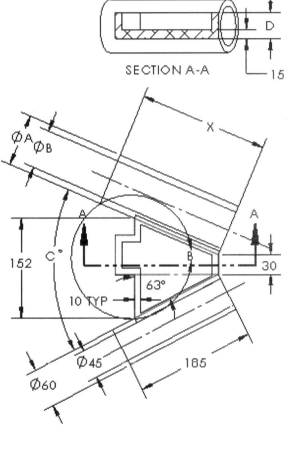

SECTION A-A

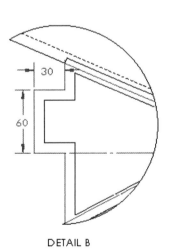

DETAIL B

Let's begin.

Compare the **dimensions**, **material**, **variables** and **equation**.

Question 1: Part II

Compare the provided information with the existing part (CSWP PP2-2) which you created in the previous section. Use the provided initial and final models with the rollback bar if needed.

What is the mass of the part in grams?

Provided information:

Units: MMGS (millimeter, gram, second)

Decimal Places: 2

Part Origin: Arbitrary

Material: 6061 Alloy

Density: 2700 kg/m^2

All holes through all unless shown otherwise.

Use the following parameters and equations which correspond to the dimensions labeled in the images.

A: 80 mm

B: 60 mm

C: 50 degree

D: 40 mm

X: 180 + D/2

Question 1: Part III

Build the following part in SOLIDWORKS.

Compare the provided information (dimensions, material, variables and equation) with the existing part (CSWP PP2-3) which you created in the previous section.

What is the mass of the part in grams?

Provided Information:

Units: MMGS (millimeter, gram, second)

Decimal Places: 2

Part Origin: Arbitrary

Material: 6061 Alloy

Density: 2700 kg/m^2

All holes through all unless shown otherwise.

Use the following parameters and equations which correspond to the dimensions labeled in the images.

A: 80 mm

B: 60 mm

C: 50 degrees

D: 40 mm

X: 180 + D/2

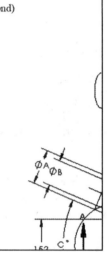

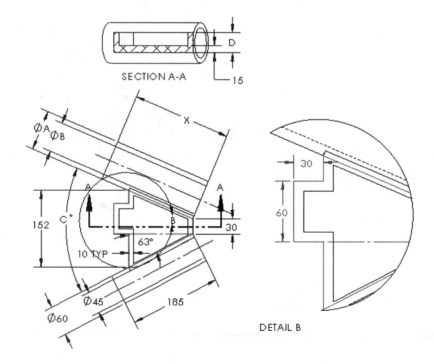

SECTION A-A

DETAIL B

None of the variables or equation was modified between the second and third question. Start with the first Boss-Extrude1 feature.

1. **Drag** the rollback bar directly under Boss-Extrude1.

2. **Edit** the base sketch (Sketch1) according to the Detail view. Create a 30mm x 60mm corner rectangle. Trim geometry.

A: 80 mm	A: 80 mm
B: 60 mm	B: 60 mm
C: 50 degree	C: 50 degree
D: 40 mm	D: 40 mm
X: 180 + D/2	X: 180 + D/2

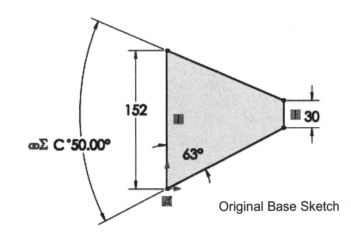

Original Base Sketch

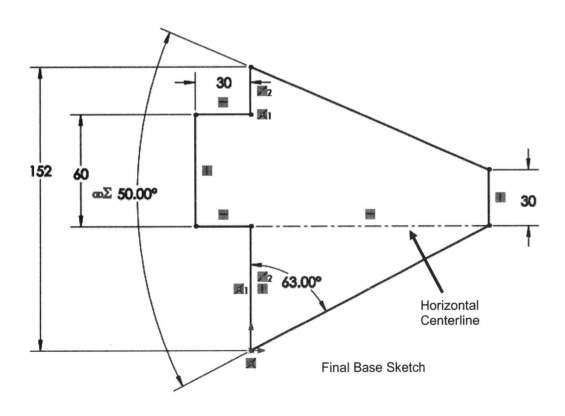

Final Base Sketch

3. **Save** the part.

4. **Calculate** the mass of the model in grams.

What is the mass of the part in grams?

5. Enter **2645.10 grams**.

🔅 Always enter the needed decimal places in the answer field.

You are finished with this section. Good luck on the exam in Segment 1.

🔅 At the time of this writing, there were slight variations in Mass Properties between SOLIDWORKS versions. These variations are still within the 1% required for the CSWP exam segments. Always save your models to verify your results.

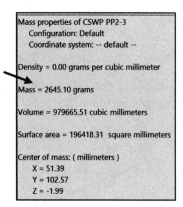

Mass properties of CSWP PP2-3
 Configuration: Default
 Coordinate system: -- default --

Density = 0.00 grams per cubic millimeter

Mass = 2645.10 grams

Volume = 979665.51 cubic millimeters

Surface area = 196418.31 square millimeters

Center of mass: (millimeters)
 X = 51.39
 Y = 102.57
 Z = -1.99

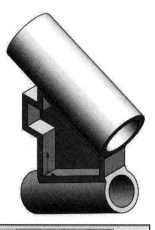

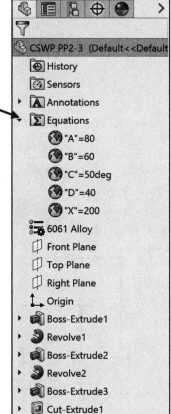

CSWP PP2-3 (Default<<Default

- 🗓 History
- 🔲 Sensors
- ▸ 🅰 Annotations
- ▾ Σ Equations
  - 🌐 "A"=80
  - 🌐 "B"=60
  - 🌐 "C"=50deg
  - 🌐 "D"=40
  - 🌐 "X"=200
- 🔧 6061 Alloy
- 🔲 Front Plane
- 🔲 Top Plane
- 🔲 Right Plane
- ⌐ Origin
- ▸ 🔩 Boss-Extrude1
- ▸ 🔩 Revolve1
- ▸ 🔩 Boss-Extrude2
- ▸ 🔩 Revolve2
- ▸ 🔩 Boss-Extrude3
- ▸ 🔩 Cut-Extrude1

Below are former screen shots from a previous CSWP exam in Segment 1.

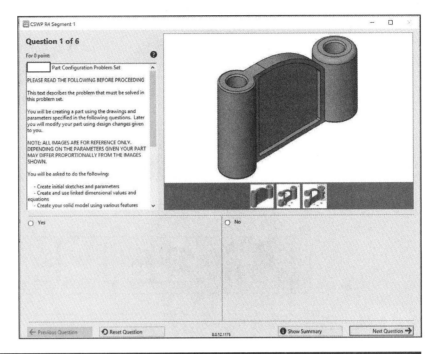

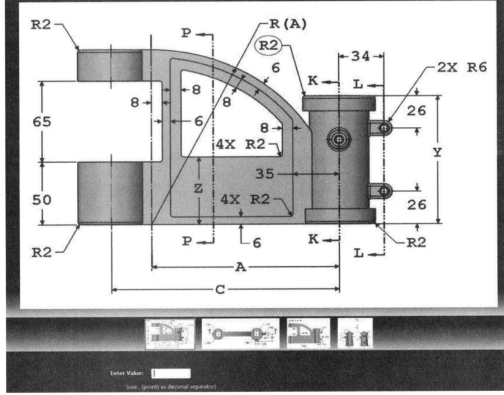

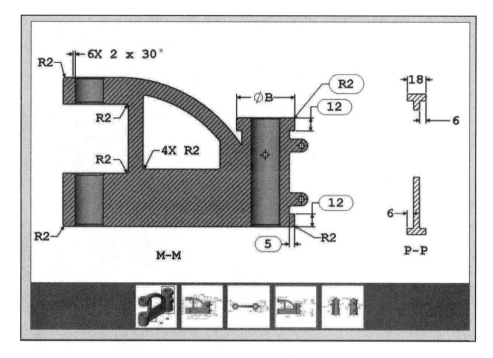

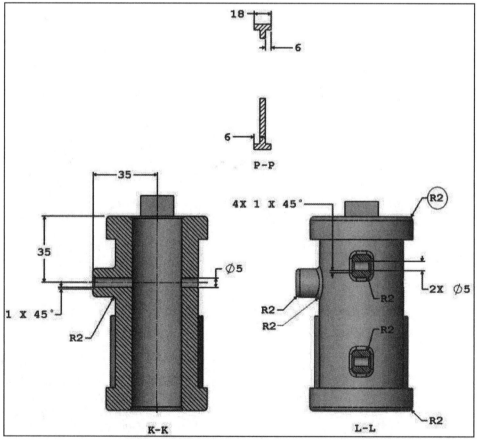

CHAPTER 2 - SEGMENT 2 OF THE CORE CSWP EXAM

Introduction

The second segment of the CSWP exam is 40 minutes with nine (9) questions (11 total but two are instructional pages) divided into two categories.

The format is either multiple choice or single answer. The first question in this segment is typically in a multiple choice format.

This section presents a representation of the types of questions that you will see in this segment of the exam.

First Category:

- Creating configurations from other configurations.

- Modifying configurations using a design table.

- Obtaining Mass Properties of various modified configurations.

Second Category:

- Modifying features and sketches of an existing part.

- Recovering from rebuild errors while maintaining the overall design intent.

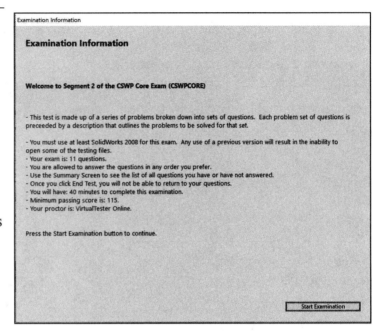

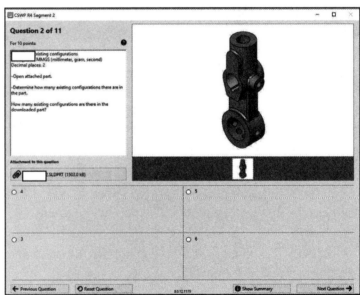

Actual CSWP exam format

You should have the exact answer (within 1% of the stated value in the multiple choice section) before you move on to the next question. If you don't have the exact answer, you will most likely fail the following question.

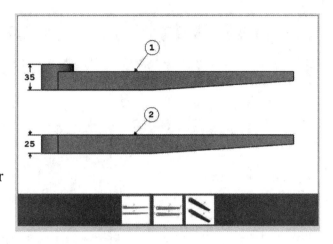

This is crucial as there is no partial credit.

A total score of 115 out of 155 or better is required to pass the second segment.

You will be tested on data found in the Mass Properties section of SOLIDWORKS. It is important to be familiar with accessing Mass Properties and interpreting them correctly.

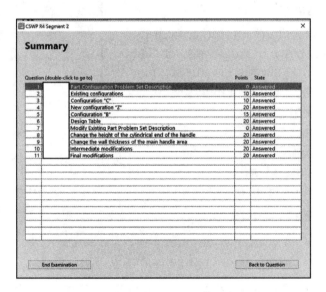

Actual CSWP exam format

SOLIDWORKS will not accept a comma in the answer field. Only use decimal points.

Use the SOLIDWORKS Tutorials and SOLIDWORKS Help to find additional information on configurations and design tables before the exam. The exam is 40 minutes; a novice using help files will not obtain a passing score.

This segment shows whether a user is capable of modifying basic features and getting needed data without wasting a lot of time.

If you fail a segment of the exam, you need to wait 14 days until you can retake that same segment. In that time, you can take another segment.

The images displayed on the exam are not to scale due to differences in the parts being downloaded for each tester.

Utilize the segment model folders provided to follow along while using the book.

Segment 2 of the CSWP CORE Exam

Load the Testing client and read the instructions. Create a folder to save your working models. Question 1 is an instructional page.

Question 2 prompts you to click a link which will download the needed file. Download the file in your created folder for the exam.

Open the Sample6 part from the Segment 2 Initial folder provided in the book. This is the downloaded part for this segment.

In this section:

- Create configurations from other configurations

- Modify configurations using a design table

- Obtain Mass Properties of various modified configurations

Let's begin.

Open the SOLIDWORKS part. A question in this section could be - determine how many existing configurations there are in the part.

1. **Create** a folder to save your working model.

2. **Open** the Sample6 part from the Segment 2 Initial folder.

3. **Click** the ConfigurationManager tab in the design tree to view the different configurations of the part.

4. **Select** 3 for the number of configurations in the multiple choice answer section of the exam.

5. **Double-click** on each configuration in the ConfigurationManager. The FeatureManager displays a different material for each configuration.

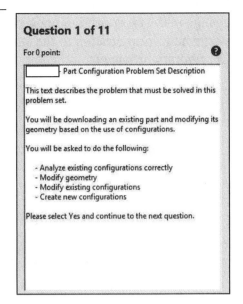

Question 1 of 11

For 0 point:

[ ] - Part Configuration Problem Set Description

This text describes the problem that must be solved in this problem set.

You will be downloading an existing part and modifying its geometry based on the use of configurations.

You will be asked to do the following:

- Analyze existing configurations correctly
- Modify geometry
- Modify existing configurations
- Create new configurations

Please select Yes and continue to the next question.

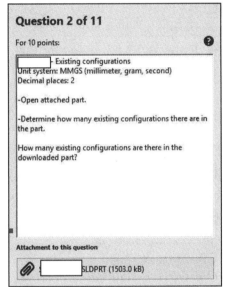

Question 2 of 11

For 10 points:

[ ] - Existing configurations
Unit system: MMGS (millimeter, gram, second)
Decimal places: 2

-Open attached part.

-Determine how many existing configurations there are in the part.

How many existing configurations are there in the downloaded part?

Attachment to this question

[📎] [ ] SLDPRT (1503.0 kB)

Actual CSWP exam format

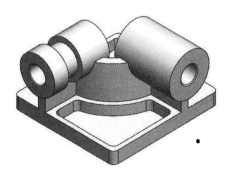

It is important that you understand that these materials will change in the exam, and to understand where you would start a new configuration to the existing ConfigurationManager.

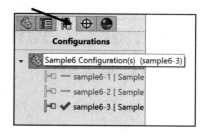

A question in this section could be - create a new configuration (sample6-4) from the existing sample6-3 configuration.

Calculate the mass of the new configuration (sample6-4) in grams. **Decimal places**: 2.

Let's begin.

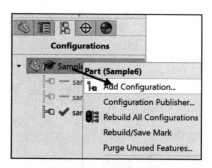

6. **Double-click** sample6-3. Sample6-3 is the active Configuration.

7. **Add** a new configuration sample6-4. Sample6-4 is the current configuration.

8. **Calculate** the mass of the new configuration sample6-4 in grams.

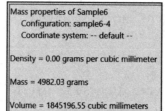

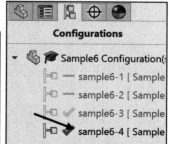

9. **Enter 4982.03** grams for the answer. It is very important to input the proper decimal places as requested in the exam. In this case, (2) two.

Mass properties of Sample6
   Configuration: sample6-4
   Coordinate system: -- default --

Density = 0.00 grams per cubic millimeter

Mass = 4982.03 grams

Volume = 1845196.55 cubic millimeters

10. **Save** the model.

A question in this section could be - suppress a feature in some but not all configurations.

Suppress the front cut (Cut-Extrude3) and back cut (Cut-Extrude5) in the sample6-4 configuration.

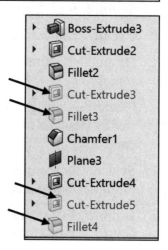

Note: The Cut-Extrude3 feature references the Fillet3, Cut-Extrude5, and Fillet4 feature in the provided model.

Calculate the mass of the part in grams.

Let's begin.

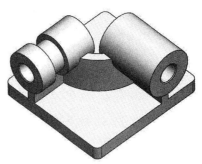

💡 There are numerous ways to address the models in this segment. A goal is to display different design intents and techniques.

11. **Suppress** the features as illustrated in the sample6-4 configuration.

12. **Calculate** the Mass Properties of configuration sample6-4 with the suppressed features.

13. **Enter 6423.17** grams for the answer. You need to be within 1% of the stated value in the single answer format to get this question correct. It is also very important to input the proper decimal places as requested in the exam. In this case there are (2) two. Round your answers if needed.

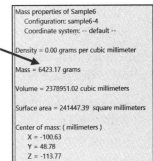

14. **Save** the part.

A question in this section could be - create configuration sample6-5 from configuration sample6-1.

Provided Information:

Add a 30x30x10mm square in the *front left corner* of (Boss-Extrude1). Utilize the existing fillet radius as the Base Extrude1 feature in your sketch.

Calculate the mass of the part in grams.

Decimal places: 2.

Let's begin.

15. **Double-click** the sample6-1 configuration from the ConfigurationManager. Sample6-1 is the active configuration.

16. **Add** a new configuration sample6-5. Sample6-5 is the current configuration.

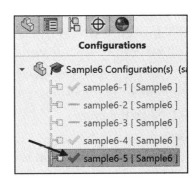

Create the sketch for the 30x30x10 mm square. Control the sketch with geometric relations and dimensions. Do not select the midpoint of the horizontal line segment.

🔅 Utilize the split bar to work between the FeatureManager and the ConfigurationManager.

17. **Create** the sketch (Sketch15) using the Corner Rectangle Sketch tool. Insert the needed geometric relation and dimension. Do not select the midpoint of the horizontal line segment.

Remember it stated, utilize the existing fillet radius as the Base Extrude1 feature in your sketch.

18. **Utilize** the Convert Entities Sketch tool along with the Trim Entitles Sketch tool to complete the sketch.

19. **Create** the Boss-Extrude feature with a depth of 10mm. Boss-Extrude4 is displayed in the FeatureManager.

20. **Calculate** the Mass Properties of configuration sample6-5 with the new feature.

21. **Enter 16500.44** grams for the answer. It is very important to input the proper decimal places as requested in the exam.

22. **Save** the part.

🔅 Utilize the segment model folders provided in the book to follow along while using the book.

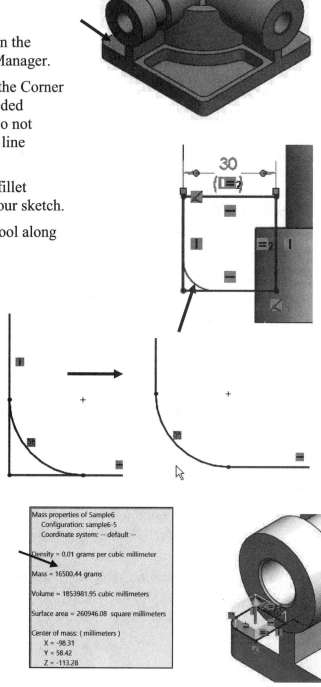

Mass properties of Sample6
   Configuration: sample6-5
   Coordinate system: -- default --

Density = 0.01 grams per cubic millimeter

Mass = 16500.44 grams

Volume = 1853981.95 cubic millimeters

Surface area = 260946.08 square millimeters

Center of mass: (millimeters)
   X = -98.31
   Y = 58.42
   Z = -113.28

A question in this section could be - create a new configuration sample6-6 from configuration sample6-2 in a design table by copying/pasting a new row.

Start with configuration sample6-2.

Create a design table.

Create a new configuration sample6-6 from configuration sample6-2 in the design table by copying /pasting a new row.

Change the following parameters in the design table for configuration sample6-6 to the following specified values:

- D1@Boss-Extrude1 = 35

- A@Sketch1 = 230

- B@Sketch1 = 225

Calculate the mass of the part in grams.

Decimal places: 2.

Let's begin.

1. **Double-click** sample6-2. Sample6-2 is the active configuration.

2. **Create** a design table. Use the Auto-create option.

3. **Create** a new configuration sample6-6 (copy/paste) from sample6-2 in the design table.

Check your new configuration in the ConfigurationManager. Update the design table.

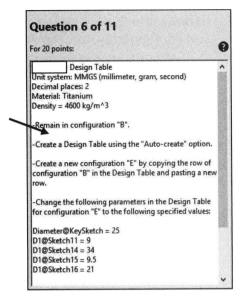

Question 6 of 11

For 20 points:

       Design Table
Unit system: MMGS (millimeter, gram, second)
Decimal places: 2
Material: Titanium
Density = 4600 kg/m^3

-Remain in configuration "B".

-Create a Design Table using the "Auto-create" option.

-Create a new configuration "E" by copying the row of configuration "B" in the Design Table and pasting a new row.

-Change the following parameters in the Design Table for configuration "E" to the following specified values:

Diameter@KeySketch = 25
D1@Sketch11 = 9
D1@Sketch14 = 34
D1@Sketch15 = 9.5
D1@Sketch16 = 21

Actual CSWP
exam format

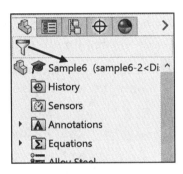

Sample6 (sample6-2<Di
  History
  Sensors
  Annotations
  Equations

Design Table

Source
○ Blank
◉ Auto-create
○ From file

Browse...

☐ Link to file

Edit Control
◉ Allow model edits to update the design table

○ Block model edits that would update the design table

4. **Exit** the design table and view the new configuration sample6-6 in the ConfigurationManager.

5. **Double-click** sample6-6. Sample6-6 is the active configuration.

6. **Edit** the design table. Do not add any columns and rows to the table at this time. View the updated design table.

Change the D1@Boss-Extrude1 = 35 parameter in the design table for configuration sample6-6.

7. **Double-click** on the Boss-Extrude1 feature. The column for the state of Boss-Extrude1 is displayed in the design table.

8. **Click** the depth dimension 25 in the Graphics window. 25 is displayed in the design table. 25 is the depth dimension for all configurations.

9. **Enter** 35, depth dimension for configuration sample6-6.

10. **Exit** the design table. View the new sample6-6 configuration dimension.

11. **Apply** the Measure tool to confirm the design table modification.

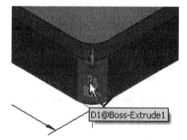

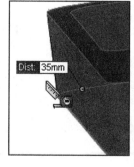

Change the following parameters in the design table for configuration sample6-6:

- A@Sketch1 = 230

- B@Sketch1 = 225

12. **Edit** the design table. Do not add any columns and rows to the table at this time. View the updated design table.

	A	B	C	D	E	F	G	H	I	J	K	L	M
1	Design Table for: Sample6												
2		$DESCRIPTION	$COLOR	X@Sketch3	Y@Sketch5	$STATE@Sketch15	$STATE@Boss-Extrude4	$STATE@Cut-Extrude3	$STATE@Fillet3	$STATE@Cut-Extrude5	$STATE@Fillet4	$STATE@Boss-Extrude1	D1@Boss-Extrude1
3	sample6-1	sample6-1	5659017	76.33333333	87.33333333	U	U	U	U	U	U	U		25
4	sample6-2	sample6-2	15266559	76.33333333	87.33333333	S	S	U	U	U	U	U		25
5	sample6-3	sample6-3	16771293	76.33333333	87.33333333	U	U	U	U	U	U	U		25
6	sample6-4	sample6-4	16771293	76.33333333	87.33333333	U	U	S	S	S	S	U		25
7	sample6-5	sample6-5	5659017	76.33333333	87.33333333	U	U	U	U	U	U	U		25
8	sample6-6	sample6-6	15266559	76.33333333	87.33333333	S	S	U	U	U	U	U		35
9													
10													
11													
12													
13													
14													
15													
16													
17													
18													
19													
20													

Sheet1

13. **Double-click** the front face of Boss-Extrude1 in the Graphics window. Dimensions Λ and B are displayed.

14. **Double-click** inside cell N2.

15. **Double-click** dimension A in the Graphics window. Click OK.

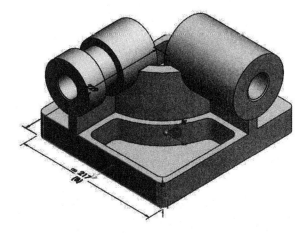

16. **Double-click** inside cell O2.

17. **Double-click** dimension B in the Graphics window. Click OK. View the updated design table.

18. **Double-click** inside cell N8.

19. **Enter** 230 as illustrated.

20. **Double-click** inside cell O8.

21. **Enter** 225 as illustrated.

	A	B	C	D	E	F	G	H	I	J	K	L	M	N	O
1	Design Table for: Sample6														
2		$DESCRIPTION	$COLOR	X@Sketch3	Y@Sketch5	$STATE@Sketch15	$STATE@Boss-Extrude4	$STATE@Cut-Extrude3	$STATE@Fillet3	$STATE@Cut-Extrude5	$STATE@Fillet4	$STATE@Boss-Extrude1	D1@Boss-Extrude1	A@Sketch1	B@Sketch1
3	sample6-1	sample6-1	5659017	76.33333333	87.33333333	U	U	U	U	U	U	U	25	229	217
4	sample6-2	sample6-2	15266559	76.33333333	87.33333333	S	S	U	U	U	U	U	25		
5	sample6-3	sample6-3	16771293	76.33333333	87.33333333	U	U	U	U	U	U	U	25		
6	sample6-4	sample6-4	16771293	76.33333333	87.33333333	U	U	S	S	S	S	U	25		
7	sample6-5	sample6-5	5659017	76.33333333	87.33333333	U	U	U	U	U	U	U	25		
8	sample6-6	sample6-6	15266559	76.33333333	87.33333333	S	S	U	U	U	U	U	35	230	225
9															
10															
11															
12															
13															
14															
15															
16															
17															
18															
19															
20															

22. **Exit** the design table.

23. **Double-click** sample6-6. Sample6-6 is the active configuration.

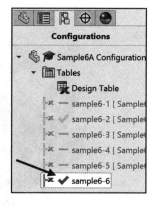

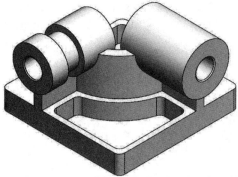

24. **Double-click** on the Boss-Extrude1 feature.

25. **View** the modified dimensions.

26. **Calculate** the Mass Properties of configuration sample6-6.

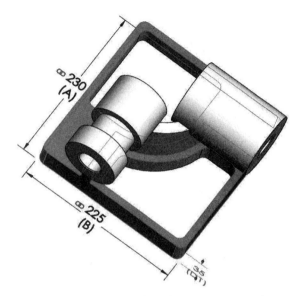

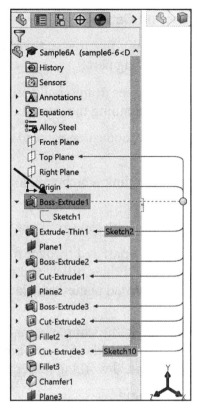

27. **Enter 16604.47** grams for the anwser. In the CSWP exam you will need to enter this number exactly. You are finished with the first category in Segment 2 of the CSWP CORE exam.

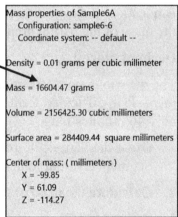

Mass properties of Sample6A
   Configuration: sample6-6
   Coordinate system: -- default --

Density = 0.01 grams per cubic millimeter

Mass = 16604.47 grams

Volume = 2156425.30 cubic millimeters

Surface area = 284409.44 square millimeters

Center of mass: (millimeters)
   X = -99.85
   Y = 61.09
   Z = -114.27

Second Category:

- Modify features and sketches of an existing part
- Recover from rebuild errors while maintaining the overall design intent

In the second category read question 7 and 8 carefully.

Click the link which will download the needed document.

Create a folder to save your working models.

Modify a part and recover from errors in this category. Edges, fillets and faces will be lost, and modified planes and sketches.

Open the sample7-1 part from the Segment 2 Initial folder. This is the downloaded working folder for this segment.

Let's begin.

A question in this section could be - determine the mass of the part in grams.

Decimal places: 2.

The questions start off simple and then increase in difficulty. Remember Segment 2 is only 40 minutes long.

1. **Open** the Sample7-1 part from the Segment 2 Initial folder.

2. **Calculate** the mass of the part in grams.

3. **Select 1443.72** grams in the multiple-choice answer section of the exam. The first question in this segment is typically in a multiple choice format.

4. **Save** Sample7-1 as Sample7-2.

A question in this section could be - modify the large hole diameter from 25mm to 20mm.

Calculate the mass of the part in grams.

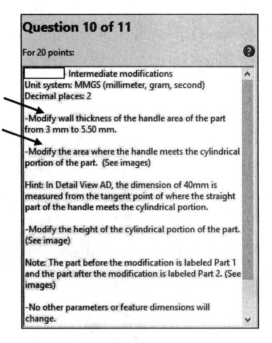

Question 10 of 11

For 20 points:

[   ]: Intermediate modifications
Unit system: MMGS (millimeter, gram, second)
Decimal places: 2

-Modify wall thickness of the handle area of the part from 3 mm to 5.50 mm.

-Modify the area where the handle meets the cylindrical portion of the part. (See images)

Hint: In Detail View AD, the dimension of 40mm is measured from the tangent point of where the straight part of the handle meets the cylindrical portion.

-Modify the height of the cylindrical portion of the part. (See image)

Note: The part before the modification is labeled Part 1 and the part after the modification is labeled Part 2. (See images)

-No other parameters or feature dimensions will change.

Actual CSWP
exam format

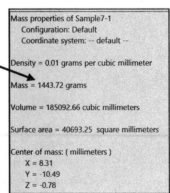

Mass properties of Sample7-1
   Configuration: Default
   Coordinate system: -- default --

Density = 0.01 grams per cubic millimeter

Mass = 1443.72 grams

Volume = 185092.66 cubic millimeters

Surface area = 40693.25 square millimeters

Center of mass: (millimeters)
   X = 8.31
   Y = -10.49
   Z = -0.78

Let's begin.

First find the large hole and its initial value.

5. **Double-click** Cut-Extrude1. View the 25mm diameter.

6. **Enter** 20mm.

7. **Rebuild** the part.

8. **Calculate** the mass of the part in grams.

9. **Enter 1522.29** grams. Think about your answer. If you obtain a lower mass number when you have more material, something is wrong.

10. **Save** Sample7-2 as Sample7-3.

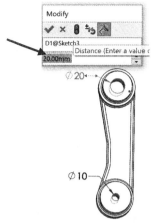

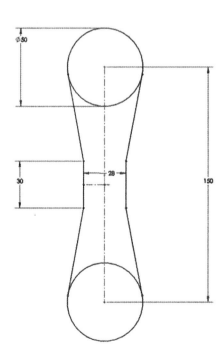

A question in this section could be - modify the shell thickness from 5mm to 4mm.

Calculate the mass of the part in grams.

Decimal places: 2.

Think about the effects of this modification on the mass of the part.

Let's begin.

11. **Edit** Shell1 from the FeatureManager.

12. **Enter** 4mm for Thickness.

13. **Calculate** the mass of the part in grams.

14. **Enter 961.25** grams.

15. **Save** Sample7-3 as Sample7-4.

A question in this section could be - create a symmetrical handle between the two cylinders.

The handle is symmetrical both horizontally and vertically.

Delete the Shell feature.

Maintain the original fillet dimension in the handle of 20mm. Maintain the overall dimension between the two cylinders of 150mm.

Calculate the mass of the part in grams.

SOLIDWORKS will present multiple views of the model in the question (an overall view of the sketch) or the overall depth of the feature.

Let's begin. Start with your Sample7-4 model.

1. **Suppress** the Shell1 feature from the FeatureManager. As a general rule, suppress features before you delete them. This will inform you if there are any rebuild or feature errors during modification in the exam.

2. **Edit** Sketch1 (Multi-body sketch) from the Boss-Extrude2 feature in the FeatureManager. View the dimensions. Keep the 150mm dimension.

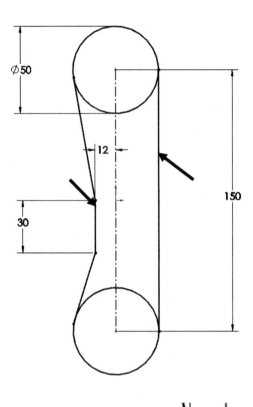

3. **Delete** any unneeded geometric relations (Horizontal) or sketch entities (vertical line) as illustrated. Deleting the Horizontal relation on the point provides the ability to move the attached sketch entities.

4. **Create** a horizontal centerline between the origin and the midpoint of the vertical line to the left of the origin.

5. **Insert** a Horizontal relation as illustrated. As a general rule, insert geometric relations before dimensions in a sketch.

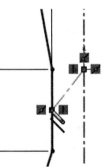

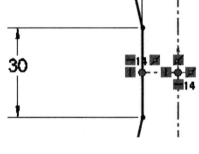

☀ There are numerous ways to address the models in this book. A goal is to display different design intents and techniques for a timed exam.

The question was to create a symmetrical handle between the two cylinders. The handle is symmetrical both horizontally and vertically. Add a symmetric relation.

6. **Select** the Centerline, the two end points on the sketch and add a symmetric relation.

Sketch1 is fully defined. Mirror the sketch to the right side.

7. **Window-select** the sketch on the left side and apply the Mirror Entities Sketch tool. Mirror about the vertical centerline. The sketch is fully defined.

The width between the two vertical lines in the initial question displayed a sketch that showed 28mm.

8. **Modify** the 12mm dimension to 14mm to obtain the needed width.

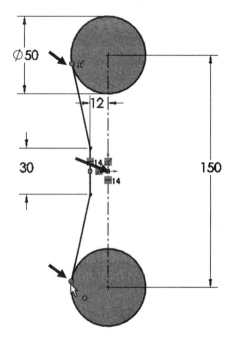

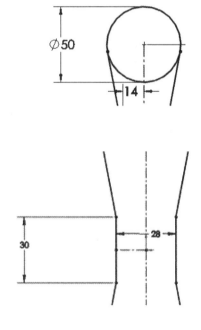

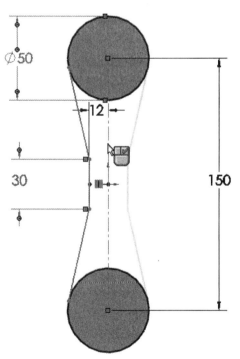

Address all needed fillet features. Maintain the original fillet dimension in the handle of 20mm.

9. **Edit** the Fillet1 feature. Select the two edges. Four edges should be displayed in the Fillet1 PropertyManager.

10. **Calculate** the mass of the part in grams.

11. **Enter 1576.76** grams.

12. **Save** Sample7-4 as Sample7-5.

SOLIDWORKS Mass Properties calculates the center of mass for every model. At every instant of time, there is a unique location (x, y, z) in space that is the average position of the systems mass. The CSWP exam asks for center of gravity. For the purpose of calculating the center of mass and center of gravity near to earth or on earth, you can assume that the center of mass and the center of gravity are the same.

Always save your models to verify your results.

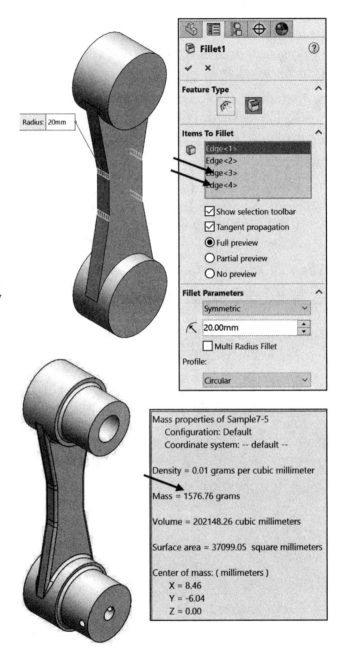

Mass properties of Sample7-5
Configuration: Default
Coordinate system: -- default --

Density = 0.01 grams per cubic millimeter

Mass = 1576.76 grams

Volume = 202148.26 cubic millimeters

Surface area = 37099.05 square millimeters

Center of mass: (millimeters)
X = 8.46
Y = -6.04
Z = 0.00

A question in this section could be - modify the depth of the Boss-Extrude3 feature.

Modify the height from 57mm to 47mm. This would mean that the height of the smaller boss is 17mm.

Delete the Shell feature.

Calculate the mass of the part in grams.

Decimal places: 2.

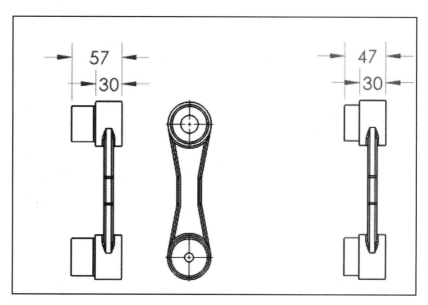

This looks like a straightforward problem. Be careful. SOLIDWORKS wants you to understand how to create features and their options.

Let's begin.

Start with the Sample7-5 model. In Sample7-5, the Shell1 feature is suppress.

1. **Edit** Boss-Extrude3 from the FeatureManager. Note the sketch is not created from the base of the larger cylinder. The Start Condition is Offset. The End Condition is Up To Surface. The two circular sketches were created on Plane1.

2. **Enter** 2mm (you need 47mm - Offset by 12 - reduce it to 2mm) in the Enter Offset Value box to obtain the needed value.

Confirm that your math is correct. Use the Measure tool during the exam.

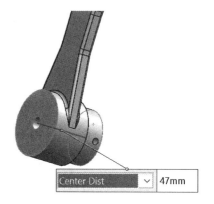

3. **Calculate** the mass of the part in grams.

4. **Enter 1401.30** grams. You need to be within 1% of the answer to get this question correct.

5. **Save** Sample7-5 as Sample7-6.

A question in this section could be - modify the sketch on the bottom cylinder to a 21mm square, centered on the cylindrical feature as illustrated.

The square boss contains no fillets.

Maintain all dimensions for both the 10mm and 5mm thru holes as illustrated.

Calculate the mass of the part in grams.

Decimal places: 2.

Think about the feature that you need to delete.

Will there be rebuild errors that you need to address? Yes, there will be in the exam.

Study the provided views and read the question carefully. Remember; there is no partial credit.

Let's begin.

Start with your model Sample7-6.

Delete the original Sketch entities. You need to modify the sketch on the bottom cylinder to a 21mm square, centered on the cylindrical feature.

6. **Edit** Sketch2 from the Boss-Extrude3 feature.

7. **Window-select** the bottom cylinder sketch entities as illustrated.

8. **Press** Delete.

9. **Click** Yes from the SOLIDWORKS dialog box.

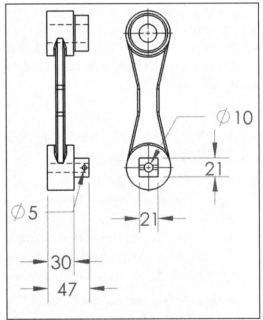

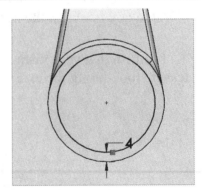

10. **Create** a Center Rectangle sketch located at the center point of the Boss-Extrude1 feature as illustrated.

11. **Insert** needed geometric relations and dimension.

You deleted a circular face to create a square face. A rebuild error is displayed. The sketch contains dimensions or relations to model geometry which no longer exists.

12. **Click** Stop and Repair.

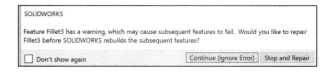

13. **Repair** the Fillet3 feature. You deleted the sketch geometry when you created the center rectangle.

14. **Repair** the Plane2 feature. Plane2 used the small cylindrical face which was deleted. Use the large cylindrical face to repair.

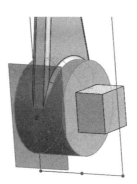

15. **Repair** the Cut-Extrude1 feature. Edit Sketch3. Delete (Coincident1) a dangling sketch entity.

16. **Insert** a Concentric relation to fully define Sketch3. Sketch3 is fully defined. Exit the sketch.

Sketch4 contains dimensions or relations to model geometry which no longer exists. Address Sketch4.

17. **Edit** the Sketch Plane of Sketch4. The original Sketch plane was located on the cylindrical face of Boss-Extrude3 which was modified. Select a new Sketch Plane.

18. A SOLIDWORKS dialog box is displayed. **Press** the Continue (Ignore Error) button.

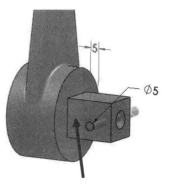

There are still a few dangling relations associated with the small hole (Cut-Extrude2).

19. **Edit** Sketch4. Delete the Coincident0 and Distance2 relation.

20. **Fully define** Sketch4. Insert the needed dimension and Coincident relation. The hole should be centered on the Boss-Extrude3 feature.

You now have a clean part with no error messages.

21. **Calculate** the mass of the part in grams.

22. **Enter 1278.37** grams.

You are finished with the Segment 2 section in this chapter. The questions in this segment are not too difficult but you only have 40 minutes for this segment. Managing your time is key in any segment of the CSWP CORE exam.

There are numerous ways to address the model in this section. A goal is to display different design intents and techniques.

Always save your models to verify the results.

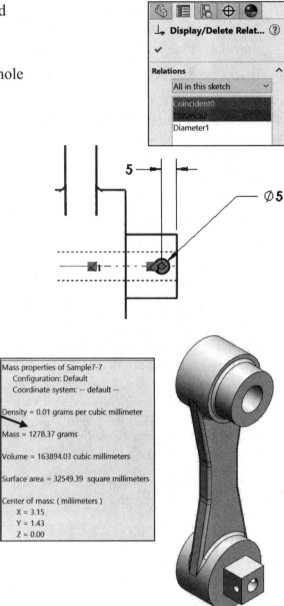

Mass properties of Sample7-7
    Configuration: Default
    Coordinate system: -- default --

Density = 0.01 grams per cubic millimeter

Mass = 1278.37 grams

Volume = 163894.03 cubic millimeters

Surface area = 32549.39 square millimeters

Center of mass: (millimeters)
    X = 3.15
    Y = 1.43
    Z = 0.00

Segment 2 - First category of the CSWP CORE exam - Additional Practice Problems

In this section, there are fewer step-by-step procedures. Utilize the rollback bar if needed in the provided initial and final models.

Question 1:

A question in this section could be - How many configurations are there associated with this part?

Let's begin.

1. **Open** the CSWP PP 4-1 part from the Segment 2 Initial folder.

2. **Click** the ConfigurationManager tab in the design tree. View the different configurations of the part.

3. **Select** 3 for the number of configurations in the multiple choice answer section of the exam. The first question in this segment is typically in a multiple choice format.

a) 1, b) 2, c) **3,** d) 4

4. **Save** the model.

💡 Double-click on each configuration in the ConfigurationManager. The FeatureManager displays a different material for each configuration. It is important that you understand that these materials will change in the exam and to understand where you would start a new configuration to the existing ConfigurationManager.

Question 1: Part II

A question in this section could be - calculate the mass in grams of configuration Example 4-2.

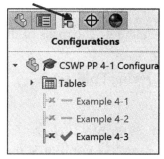

Provided Information:

Decimal places: 2.

Let's begin.

1. **Double-click** Example 4-2. Example 4-2 is the active configuration.

2. **Calculate** the mass of the Example 4-2 configuration in grams.

3. **Enter 1229.65** grams for the answer. It is very important to input the proper decimal places as requested in the exam. In this case (2) two.

4. **Save** the model.

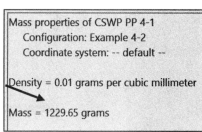

Mass properties of CSWP PP 4-1
Configuration: Example 4-2
Coordinate system: -- default --

Density = 0.01 grams per cubic millimeter

Mass = 1229.65 grams

Question 3:

A question in this section could be - create a new configuration named Example 4-4 based on the Example 4-3 configuration.

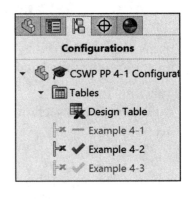

Modify the design table with the specified values.

- D1@Sketch1=70

- D2@Sketch1=50

- D3@Sketch1=46

- D5@Extrude-Thin1=30

- D1@Shell1=8

- D1@Sketch2=12

- Cut-Extrude3=Suppressed

- Cut-Extrude4=Unsuppressed

Calculate the mass of the part (Example 4-4 configuration) in grams.

Decimal places: 2.

Let's begin.

1. **Double-click** the Example 4-3 configuration. This is the active configuration.

2. **Add** a new Configuration, named Example 4-4.

3. **Edit** the Design Table. Modify the design table with the new specified values.

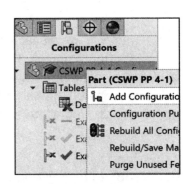

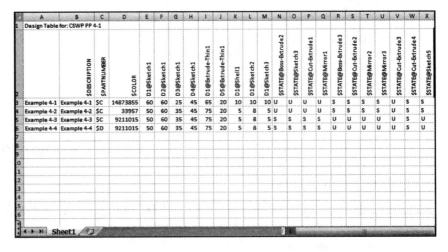

	$DESCRIPTION	$PARTNUMBER	$COLOR	D1@Sketch1	D2@Sketch1	D3@Sketch1	D4@Sketch1	D1@Extrude-Thin1	D5@Extrude-Thin1	D1@Shell1	D1@Sketch2	D1@Sketch3	$STATE@Boss-Extrude2	$STATE@Sketch3	$STATE@Cut-Extrude1	$STATE@Mirror1	$STATE@Boss-Extrude3	$STATE@Cut-Extrude2	$STATE@Mirror2	$STATE@Mirror3	$STATE@Cut-Extrude3	$STATE@Cut-Extrude4	$STATE@Sketch5
Design Table for: CSWP PP 4-1																							
Example 4-1	Example 4-1	$C	14873855	60	60	25	45	65	20	10	10	10	U	U	U	U	S	S	S	S	U	S	S
Example 4-2	Example 4-2	$C	33957	50	60	35	45	75	20	5	8	5	U	U	U	U	S	S	S	S	U	S	S
Example 4-3	Example 4-3	$C	9211015	50	60	35	45	75	20	5	8	5	S	S	S	S	U	U	U	U	U	U	U
Example 4-4	Example 4-4	$C	9211015	70	50	46	45	75	30	8	12	5	S	S	S	S	U	U	U	U	S	U	U

Sheet1

4. **Exit** the Design Table.

5. **Calculate** the mass of the part (Example 4-4 configuration) in grams.

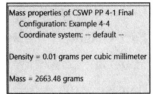

Mass properties of CSWP PP 4-1 Final
Configuration: Example 4-4
Coordinate system: -- default --

Density = 0.01 grams per cubic millimeter

Mass = 2663.48 grams

6. **Enter 2663.48** grams for the answer. It is also very important to input the proper decimal places as requested in the exam. In this case there are (2) two. Round your answers if needed.

7. **Save** the part.

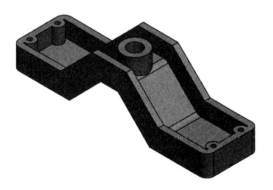

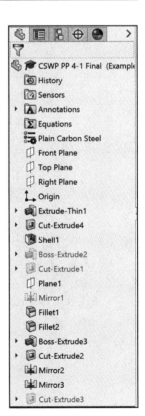

Segment 2 - Second category of the CSWP CORE exam - Additional Practice Problems

In this section, there are fewer step-by-step procedures. Utilize the rollback bar if needed in the provided initial and final models.

Question 1:

A question in this section could be - modify part CSWP PP5-1 per the provided drawing information.

What is the mass of the new part?

All holes through all unless shown otherwise.

Units: MMGS (millimeters, grams, second)

Decimal Places: 2

Part Origin: Arbitrary

Material: Cast Alloy Steel

Density: 7300 kg/m^2

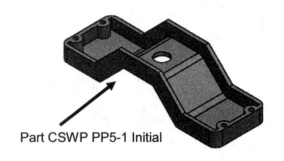

Part CSWP PP5-1 Initial

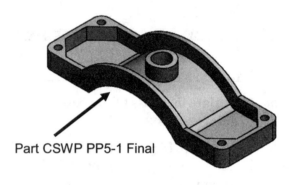

Part CSWP PP5-1 Final

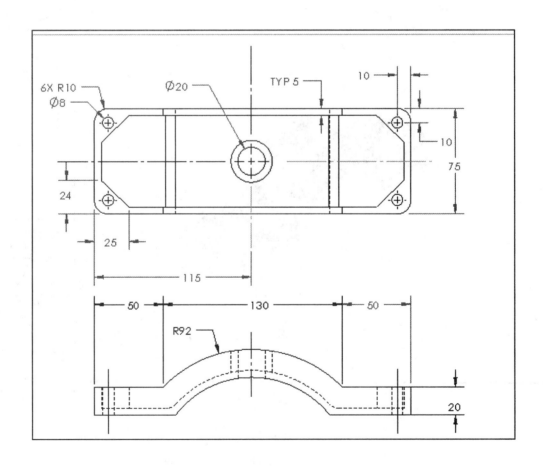

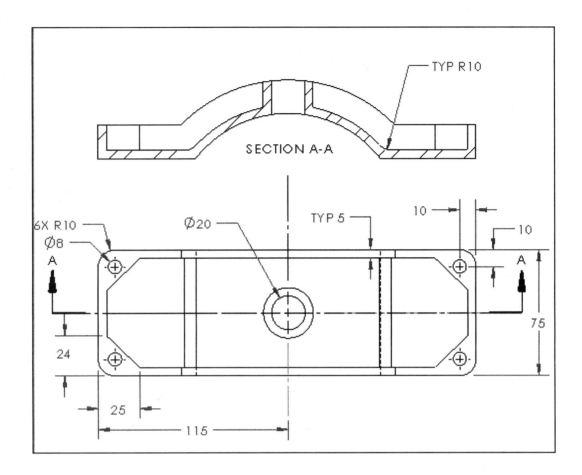

Let's begin.

1. **Create** a folder to save your working model.

2. **Open** the CSWP PP5-1 part from the Segment 2 Initial folder.

3. **View** the FeatureManager. Use the Rollbar bar to examine and modify features and sketches from the provided drawing information.

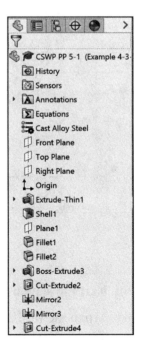

4. **Modify** the Extrude-Thin1 sketch as illustrated using a 3point Arc Sketch tool. Both sketches are fully defined. Dimensions and geometric relations define the sketches. The drawing views provide the needed information for sketch and feature modification of the part.

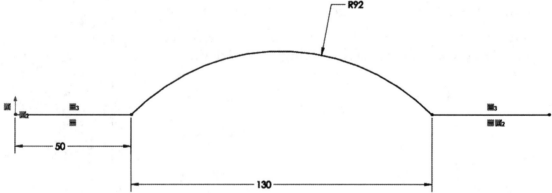

5. **Exit** the sketch. Continue and ignore any errors.

6. **Move** the rollback bar down below the Shell1 feature. Shell1 displays an error in the FeatureManager.

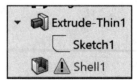

7. **Edit** the Shell feature (Shell1). Remove the missing faces and select the face of the arc to address the error.

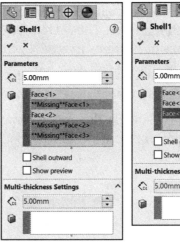

8. **Move** the rollback bar down below the Fillet2 feature. Fillet2 displays an error in the FeatureManager.

9. **Edit** the second Fillet feature (Fillet2). Remove the missing edges and update the edges where the arc and the lines meet as illustrated.

Radius: 10mm

10. **Move** the rollback bar down below the Cut-Extrude4 feature in the FeatureManager. The sketch for Cut-Extrude4 contains dimensions or relations to model geometry which no longer exist.

11. **Edit** the Sketch plane for the Cut-Extrude4 feature. There are missing items in the feature. Remove the missing plane.

12. **Select** Top Plane for the sketch plane. Remember the provided drawing information. An Extruded-Cut feature is needed on the top of the arc.

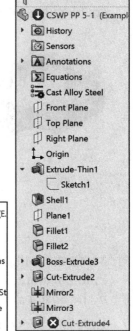

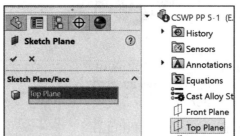

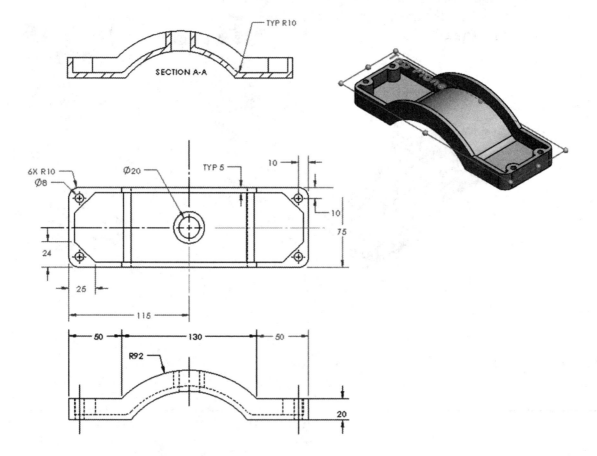

13. **Continue** and ignore any errors. **Close** the What's Wrong dialog box.

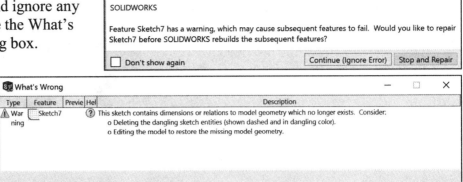

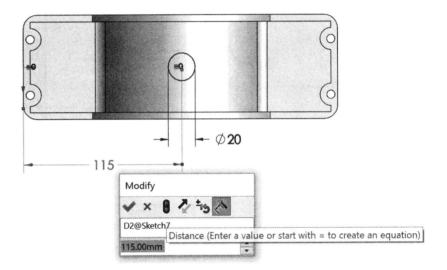

14. **Center** the hole on the part as illustrated. Delete the midpoint relation and add a dimension of 115mm from the center of the hole to the side edge of the part.

15. **Move** the Cut-Extrude4 feature before the Shell1 feature in the FeatureManager as illustrated.

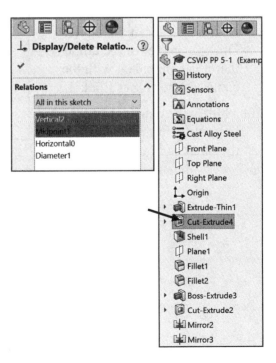

16. **View** the results in the Graphics window.

17. **Edit** the Boss-Extrude3 sketch (Sketch4). Modify the sketch per the provided information in the drawing.

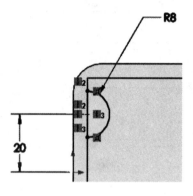

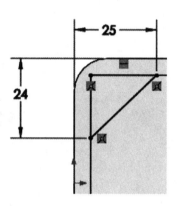

18. **Continue** and ignore any errors. Close the What's Wrong dialog box.

19. **Edit** Sketch5. Modify the location of the center hole. Delete the Coincident relation with the center of the circle. Drag the center of the circle so that it is Coincident with the center of the corner fillet.

20. **Modify** the diameter of the hole from 6mm to 8mm. Exit the Sketch.

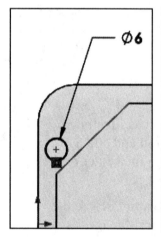

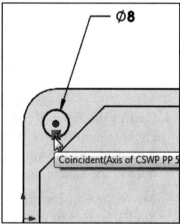

21. **Calculate** the mass of the part in grams.

22. **Enter** 1088.85 grams.

23. **Save** the part.

You are finished with this section. Good luck on the exam.

There are numerous ways to build the model in this section. A goal is to display different design intents and techniques.

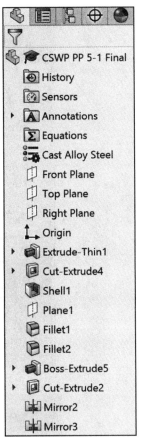

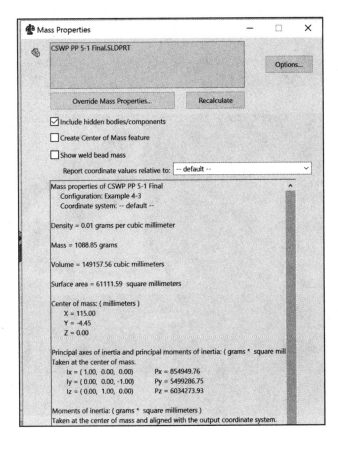

Below are former screen shots from a previous CSWP exam in Segment 2.

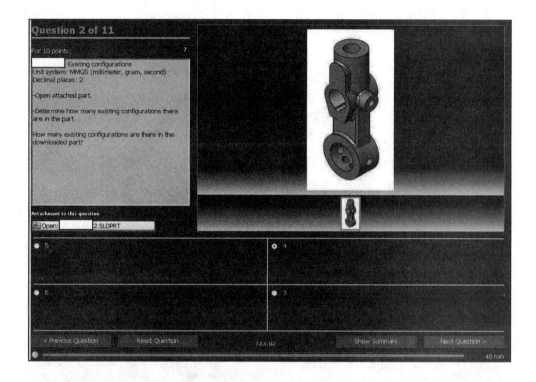

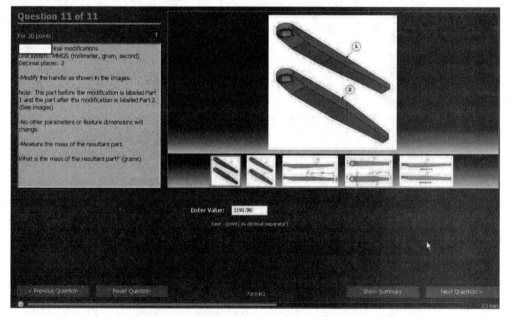

Actual CSWP exam format

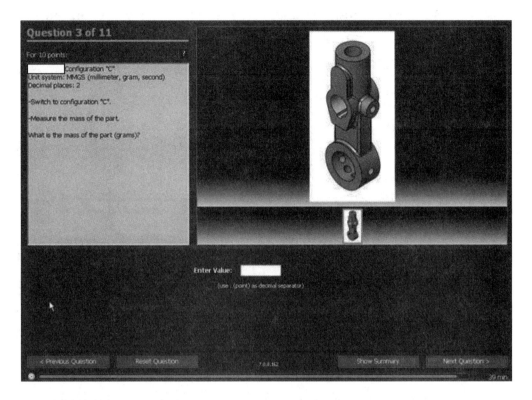

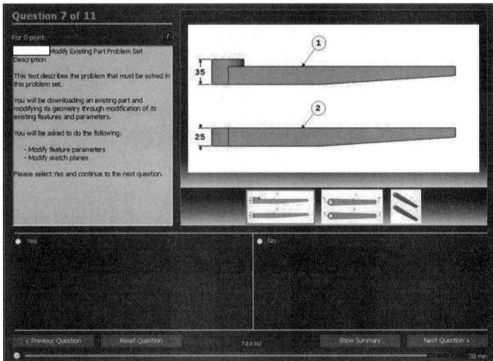

Actual CSWP exam format

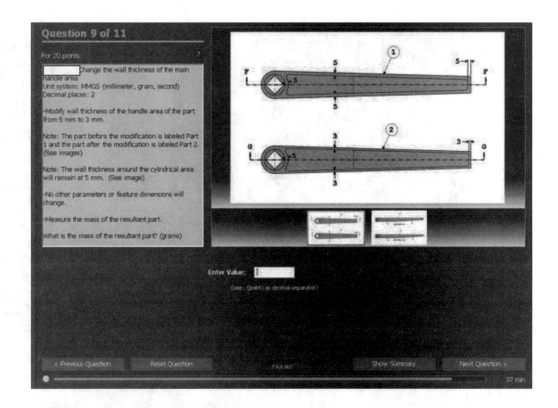

Question 9 of 11

For 20 points:

☐ Change the wall thickness of the main handle area
Unit system: MMGS (millimeter, gram, second)
Decimal places: 2

-Modify wall thickness of the handle area of the part from 5 mm to 3 mm.

Note: The part before the modification is labeled Part 1 and the part after the modification is labeled Part 2. (See images)

Note: The wall thickness around the cylindrical area will remain at 5 mm. (See image)

-No other parameters or feature dimensions will change.

-Measure the mass of the resultant part.

What is the mass of the resultant part? (grams)

Enter Value: ☐

(use . (point) as decimal separator)

< Previous Question | Reset Question | 7.0.0.162 | Show Summary | Next Question >

37 min

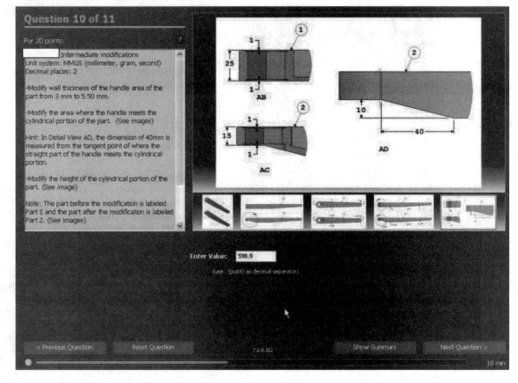

Question 10 of 11

For 20 points:

☐ Intermediate modifications
Unit system: MMGS (millimeter, gram, second)
Decimal places: 2

-Modify wall thickness of the handle area of the part from 3 mm to 5.50 mm.

-Modify the area where the handle meets the cylindrical portion of the part. (See images)

Hint: In Detail View AD, the dimension of 40mm is measured from the tangent point of where the straight part of the handle meets the cylindrical portion.

-Modify the height of the cylindrical portion of the part. (See image)

Note: The part before the modification is labeled Part 1 and the part after the modification is labeled Part 2. (See images)

Enter Value: 998.9

(use . (point) as decimal separator)

< Previous Question | Reset Question | 7.0.0.162 | Show Summary | Next Question >

18 min

CHAPTER 3 - SEGMENT 3 OF THE CSWP CORE EXAM

Introduction

The third segment is 80 minutes with twelve (12) questions (13 total but one is an instructional page). The format is either multiple choice or single fill in the blank. The first question in this segment is typically in a multiple choice format.

This section presents a representation of the types of questions that you will see in this segment of the exam.

☀ Utilize the segment model folders to follow along while using the book.

Questions 2 through 13 ask you to perform the following general tasks:

- Create a simple component and sub-assembly.

- Download parts and assemble components and a sub-assembly into an assembly.

- Insert Width, Angle, Parallel, Distance, and other mate types.

- Determine center of gravity based on a unique reference coordinate system and total mass of the assembly.

- Modify mates.

- Replace components in the assembly.

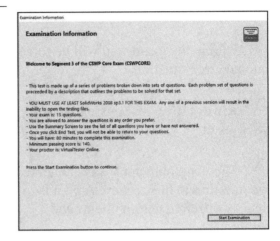

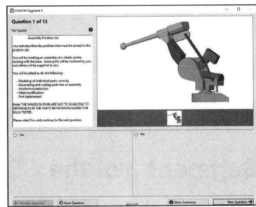

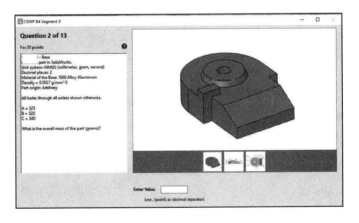

Actual CSWP exam format

- Calculate interference detection and measure distances and angles.

A total score of 140 out of 190 or better is required to pass the third segment.

Utilize the model folder to follow along while using the book.

💡 During the exam, you will be forced into an error situation.

💡 Always save your models to verify your results.

💡 There are numerous ways to address the question in this section. A goal is to display different design intents and techniques.

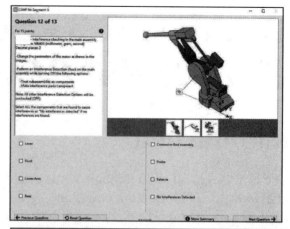

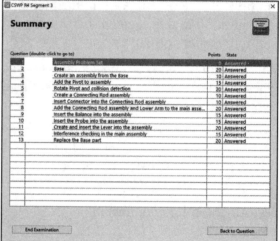

Segment 3 of the CSWP Core Exam

Create an assembly (Welding Arm) starting with the Base1 component. Some components in the exam are modeled by you, and others will be supplied to you.

Load the Testing client and read the instructions. Create a folder to save your working models.

The first question in this segment is instructional as illustrated.

The second question can be - build the Base1 component for the assembly.

Calculate the mass of the Base1 part in grams.

Information on the Base1 part is provided to you in a variety of drawing views (similar to the Segment 1 CSWP CORE) exam.

Provided information:

Unit system: MMGS (millimeter, gram, second)

Decimal place: 2

Material of Base1: Cast Stainless Steel

Density: 0.0077 g/m^3

Part origin: Arbitrary

All holes through all unless shown otherwise.

Note: Base1 is displayed in an Isometric view.

View 1:

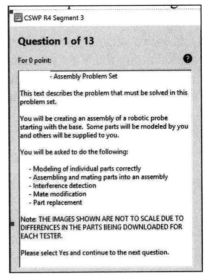

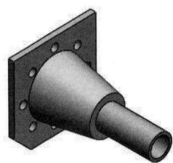

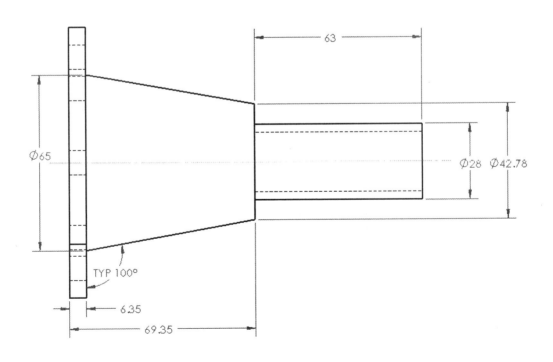

Base1

View 2:

Details are very important. Note the location of the top seed feature (Cut-Extrude1) located along the Y-axis in the drawing view. The Chamfer feature is located in the lower corner. This is important when you insert the Base1 component into the assembly document to calculate the center of gravity relative to the specified coordinate system location.

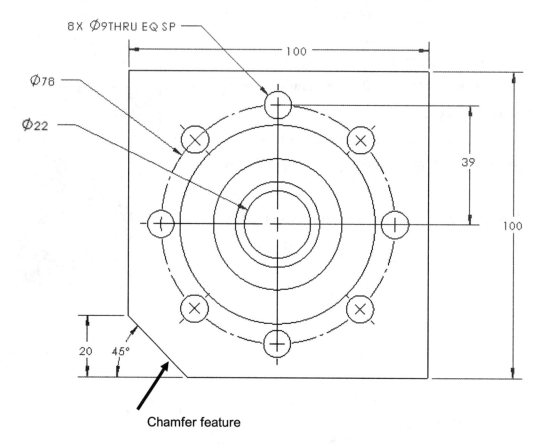

Chamfer feature

Take your time to first identify the drawing views and to better understand the provided geometry that is needed to create the initial part.

☀ There are numerous ways to build this model.

☀ The images displayed on the exam are not to scale due to differences in the parts being downloaded for each tester.

☀ SOLIDWORKS will provide a part to create that is not orientated correctly in the assembly. Knowledge of component orientation in an assembly is required along with creating a new coordinate system.

Let's begin.

Create the Base1 part. View the provided model and drawing information for dimensions and material type. Set document units and precision. The first component in an assembly should be fixed to the origin or fully defined.

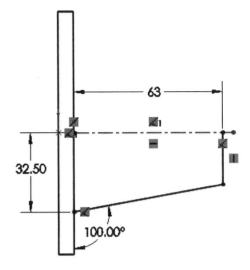

1. **Create** a folder to save your working models.

2. **Set** document properties (drafting standard, units and precision) for the model.

3. **Create** Sketch1 for Base1 on the Right Plane per the illustrated Isometric view.

4. **Create** the Extruded Base (6.35 Depth) feature.

5. **Apply** material type, Cast Stainless Steel.

6. **Create** the Revolve1 feature using Sketch2 on the Front Plane (69.35mm - 6.35mm).

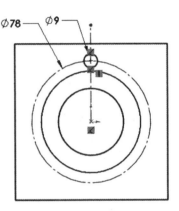

7. **Create** the first Extruded Cut feature (seed) for the circular pattern.

8. **Create** the Circular pattern (8 Instances) feature.

9. **Create** the Boss-Extrude feature (63mm Depth) with the needed sketch relations and dimensions.

10. **Create** the Chamfer feature.

11. **Calculate** the mass in grams.

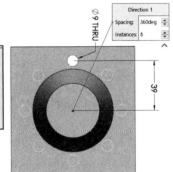

12. **Select 1690.78** grams in the multiple-choice answer section of the exam. The first question in this segment is typically in a multiple choice format.

Mass properties of Base1
   Configuration: Default
   Coordinate system: -- default --

Density = 0.01 grams per cubic millimeter

Mass = 1690.78 grams

13. **Save** the part. Name it Base1.

It is good practice to save frequently and to rename the part or assembly if you need to go back during the exam.

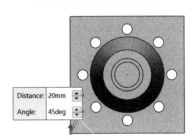

💡 You should have the exact answer (within 1% of the stated value in the multiple choice section) before you move on to the next question. If you don't have the exact answer of the first question, you will most likely fail the following question. This is crucial as there is no partial credit.

A question in this segment could be - create an Assembly document from the Base1 component.

Calculate the center of gravity in mm of the assembly relative to the new coordinate system (Coordinate System1).

Provided Information:

Unit system: MMGS (millimeter, gram, second)

Decimal places: 2

Assembly origin: Arbitrary

Orientate the Base1 component as illustrated. Create a coordinate system in the lower left corner vertex of the Base1 component as illustrated. Use this coordinate system throughout the problem set.

💡 SOLIDWORKS will provide a part to create that is not orientated correctly in the assembly. Knowledge of component orientation in an assembly is required along with creating coordinate systems.

Let's begin.

1. **Create** an assembly document and insert the Base1 part. By default the Base1 component is fixed to the origin.

There are numerous ways to modify the orientation of a component in an assembly document that is fixed to the origin.

Use the Float and Mate tools in the next section. Re-orientate the Base1 component as needed. Fully define the component in the Assembly document.

2. **Float** the part. The part is free to translate or rotate in the Assembly document (six degrees of freedom).

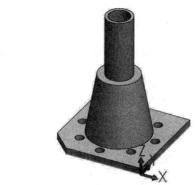

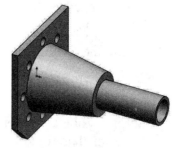

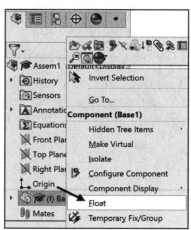

3. **Re-orientate** the assembly as needed. Create a Coincident mate between the Top Plane of the Assembly and the top face of the Base1 component.

4. **Click** the Aligned option.

The Chamfer feature is located in the wrong corner.

5. **Create** a Coincident mate between the Front Plane of the Assembly and the Top Plane of the Base1 component. The Base1 component is able to translate along the X axis.

Another mate is needed to fully define the component.

6. **Create** a Coincident mate between the Right Plane of the Assembly and the Front Plane of the Base1 component. The Base1 component is fully defined in the Assembly document with three Coincident mates.

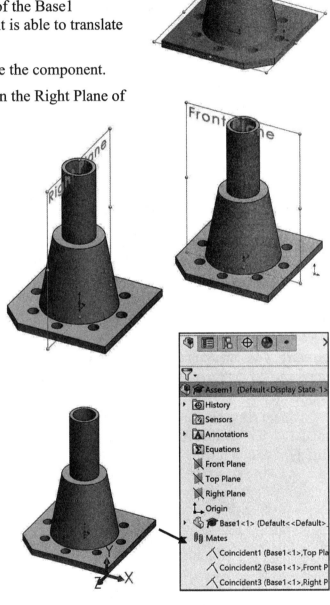

You can also rotate the Base1 component in the assembly using the Rotate Component tool (about Y & X).

Create the coordinate system to calculate the center of mass for the assembly.

7. **Click** the Coordinate System tool from the Reference Geometry drop-down menu. The Coordinate System PropertyManager is displayed.

8. **Click** the lower front right vertex for the origin.

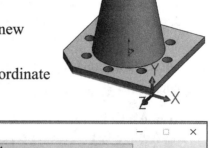

9. **Click** the bottom front edge for the X axis.

10. **Click** the bottom right edge for the Y axis. View the new coordinate system.

11. **Calculate** the center of gravity relative to the new coordinate system (Coordinate System1).

12. **Enter** the center of gravity (mm) in the three blank fields. You need to be within 1% of the answer to get this question correct.

   X = -49.75

   Y = 50.25

   Z = 29.96

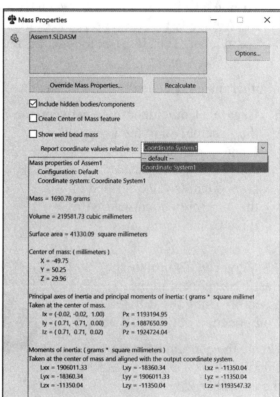

13. **Save** the assembly. In this example name the assembly Welding Arm.

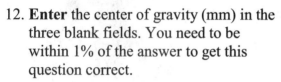 SOLIDWORKS Mass Properties calculates the center of mass for every model. At every instant of time, there is a unique location (x, y, z) in space that is the average position of the systems mass. The CSWP exam asks for center of gravity. For the purpose of calculating the center of mass and center of gravity near to earth or on earth, you can assume that the center of mass and the center of gravity are the same.

There are numerous ways to build the model in this section. A goal is to display different design intents and techniques.

A question in this segment could be - download and open the attached assemblies.

Insert them into the Welding Arm assembly. Position the assemblies (Arm and TopFixture-A) with respect to the Base1 component as illustrated.

Create all required mates.

Calculate the center of gravity in mm of the Welding Arm assembly utilizing the created coordinate system (Coordinate System1).

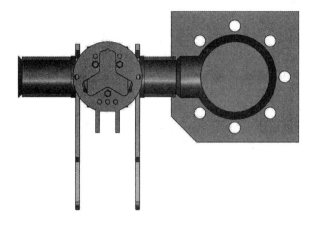

☀ Imported components have imported geometry. Select "No" on Feature Recognition. Import the geometry as quickly as possible.

Provided Information:

Unit system: MMGS (millimeter, gram, second)

Decimal places: 2

SOLIDWORKS provides symbols to indicate a Parallel mate, Width mate, Coincident mate, etc. in the exam. Take your time to review the drawing views and to understand the required mates between each component.

In this example a Parallel mate is needed between the Front Plane of the Arm sub-assembly and the front narrow face of the Base1 component.

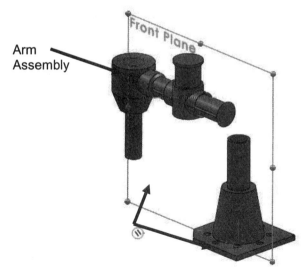

Arm Assembly

Front Plane

Select your mate and you are finished. This new mate behavior is similar to the way you add sketch relations, in that it only presents mates that are valid for the items selected.

A Parallel mate is also needed between the right flat face of the TopFixture-A assembly and the right narrow face of the Base1 component.

Let's begin.

Insert the Arm assembly. Insert two Concentric mates to have the Arm free to rotate about the Base1 component.

1. **Insert** the Arm assembly (from the Segment 3 Initial/Initial 2 folder) into the Welding Arm assembly.

2. **Insert** a Concentric mate between the outside cylindrical face of the Arm/base2-1 and the inside cylindrical face of Base1.

3. **Insert** a Coincident mate between the bottom inside flat face of Base1 and the bottom face of Arm.

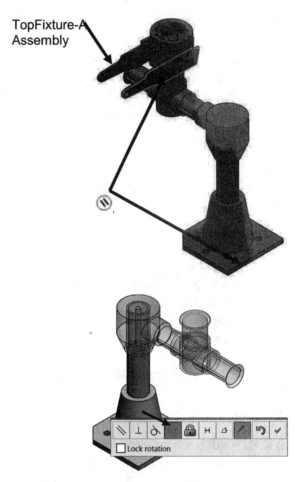

TopFixture-A Assembly

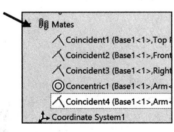

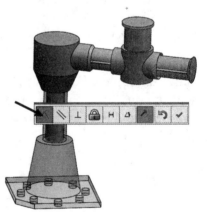

The Arm is free to rotate about the Base1 component. Next orientate the Arm parallel with the front narrow face of the Base1 component. Planes will be used in this section.

4. **Insert** a Parallel mate between the Front Plane of the Arm and the front narrow face of Base1.

5. **Click** the Anti-Aligned option for proper position if needed.

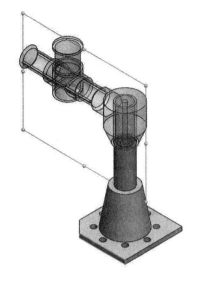

Next insert the TopFixture-A assembly into the Welding Arm assembly.

6. **Insert** the TopFixture-A assembly (from the Segment 3 Initial/Initial 2 folder) into the Welding Arm assembly.

7. **Insert** a Coincident mate between the bottom face of TopFixture-A and the top face of Arm as illustrated.

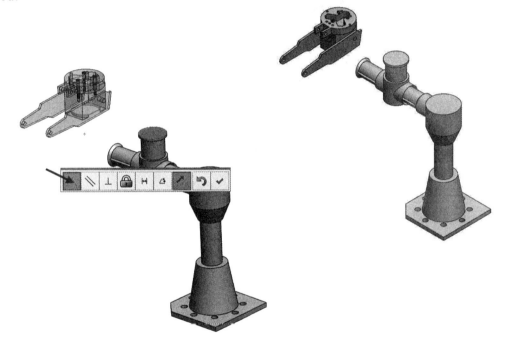

8. **Insert** a Concentric mate between the cylindrical face of the Arm and the cylindrical face of TopFixture-A as illustrated. TopFixture-A is free to rotate in the Welding Arm assembly.

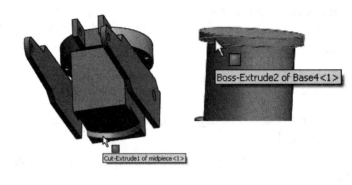

9. **Insert** a Parallel mate between the right flat face of the TopFixture-A and the right narrow face of Base1.

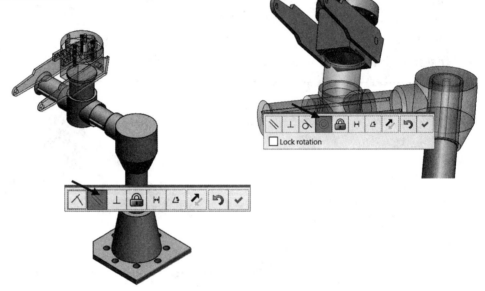

10. **Calculate** the center of gravity (mm) relative to the new coordinate system.

11. **Enter** the center of gravity (mm) in the three blank fields. You need to be within 1% of the answer to get this question correct.

X = -87.64

Y = 49.06

Z = 128.06

12. **Save** the assembly.

SOLIDWORKS Mass Properties calculates the center of mass for every model. At every instant of time, there is a unique location (x, y, z) in space that is the average position of the systems mass.

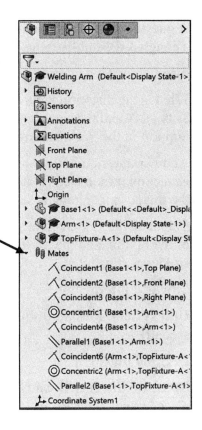

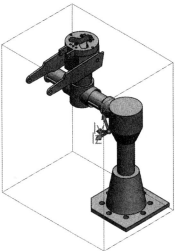

A question in this segment could be - insert the Holder-Thongs-A assembly in the Welding Arm assembly.

Insert all needed mates.

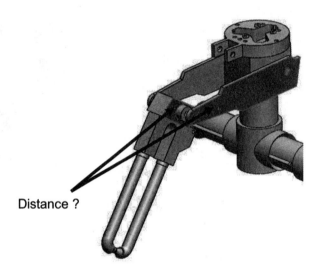

What is the distance between the right flat face of Holder-Thongs-A and the right flat face of the TopFixture-A Support in the Welding Arm assembly in mm?

Another question in this segment could be - with Collision Detection turned ON, rotate Holder-Thong-A as shown in the image until Thong is stopped by the Arm.

Calculate the angle between the flat face of Boss-Extrude1 of the Fixture and the flat face of Boss-Extrude1 of the Holder as illustrated.

Provided Information:

Unit system: MMGS (millimeter, gram, second)

Decimal places: 2

Note: Modify or delete any previously placed mates if necessary.

 Knowledge of Collision Detection and the Measure tool is required for this section.

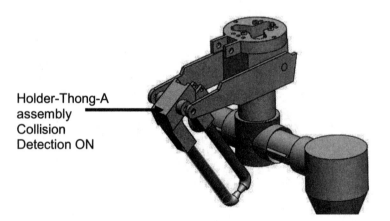

Distance ?

Holder-Thong-A assembly Collision Detection ON

Utilize the model folders to follow along while using the book.

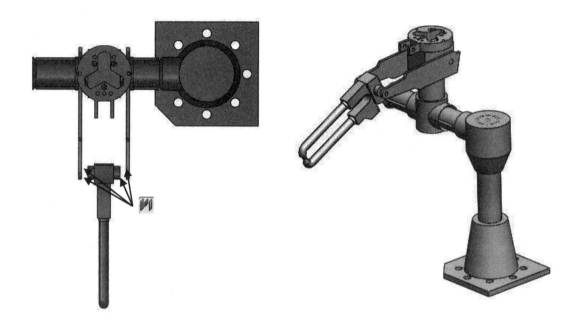

Let's begin.

Insert the Holder-Thongs-A assembly. Insert a Concentric mate and a Width mate between the Holder-Thongs-A assembly and the TopFixture assembly.

1. **Insert** the Holder-Thongs-A assembly (from the Segment 3 Initial/Initial 3 folder) into the Welding Arm assembly.

2. **Insert** a Concentric mate between the inside cylindrical face of Holder-Thongs-A and the inside cylindrical face of the TopFixture.

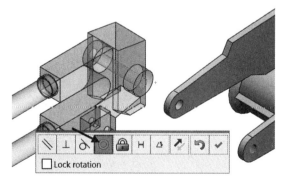

3. **Insert** a Width mate between Holder-Thongs-A and TopFixture. The Holder is free to rotate.

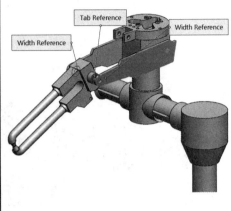

4. **Calculate** the distance between the right flat face of Holder-Thongs-A and the right flat face of TopFixture-A Support.

5. **Enter 15.48mm** for distance. You need to be within 1% of the answer to get this question correct.

6. **Save** the assembly.

You may also be required to measure angles between flat surfaces. Be certain to understand the direction, complement or supplement of the required angle.

To address the second part of the question - with Collision Detection turned ON, rotate Holder-Thong-A as illustrated until Thong is stopped by Arm.

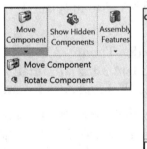

Calculate the angle between the flat face of Boss-Extrude1 of the Fixture and flat face of Boss-Extrude1 of the Holder.

7. **Activate** Collision detection from the Rotate Component PropertyManager.

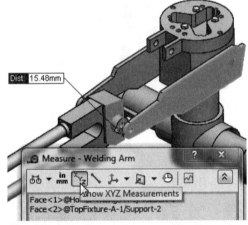

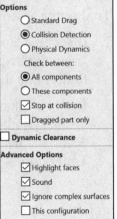

8. **Rotate** Holder-Thong-A until Thong is stopped by the Arm assembly.

9. **Calculate** the angle between the flat face of Boss-Extrude1 of the Fixture and flat face of Boss-Extrude1 of the Holder.

10. **Enter 144.84**. It is important to input the proper decimal places as requested in the exam. In this case, two (2).

11. **Save** the assembly. Note at this time the mass of the assembly is 4557.90 grams.

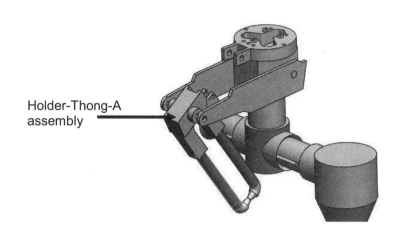

Holder-Thong-A assembly

If you don't find your answer (within 1%) in the multiple choice single answer format section, recheck your solid model for precision and accuracy.

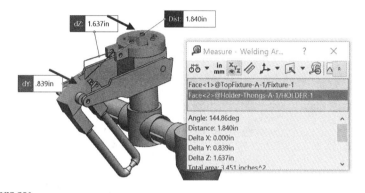

A question in this segment could be - position the Holder-Thongs-A assembly so that it interferes with the TopFixture-A assembly.

Perform an Interference Detection check on the Welding Arm assembly while turning ON the following options:

Treat sub-assemblies as components. Make interfering parts transparent.

Note: All other Interference Detection Options will be unchecked (OFF).

Select ALL the components that are found to cause interferences or "No interferences detected" if no interferences are found.

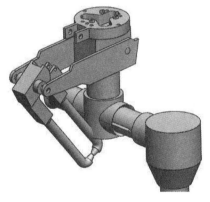

12. **Activate** the Interference Detection tool. View the PropertyManager. Clear all selections. Select the entire Welding Arm assembly.

13. **Rotate** the Holder so that it interferes with the TopFixture.

14. **Calculate** the results.

Depending on the angle of rotation of the Holder-Thongs-Assembly, the components that interfere are base2, thong, Arm, and Holder-Thongs-A. In an assembly, the location of the components that are not fully constrained will determine the volume of interference.

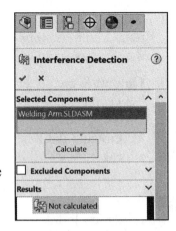

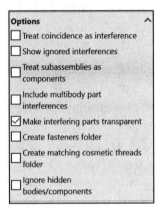

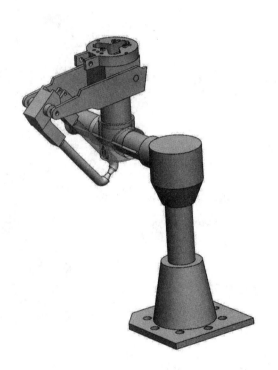

A question in this segment could be - download and open the attached components.

Create the Hydraulic and Brace Assembly.

The Hydraulic and Brace Assembly consist of two components: Hydraulic1 & BRACE.

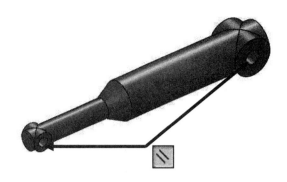

Insert a 20mm Distance mate between the bottom cylindrical face of BRACE and the bottom cylindrical face of Hydraulic1.

Calculate the mass of the assembly in grams. Save the assembly.

Later insert the assembly into the Welder Arm assembly.

Provided Information:

Unit system: MMGS (millimeter, gram, second)

Decimal places: 2

Material of components: Cast Stainless Steel

Density = 0.0077 g/m^3

Part origin: Arbitrary

 Utilize the model folders to follow along while using the book.

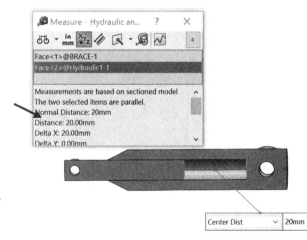

Let's begin.

1. **Create** the Hydraulic and Brace Assembly document.

2. **Set** document properties (drafting standard, units and precision).

3. **Insert** the Hydraulic1 component (Segment 3 Initial/Initial 4 folder). The Hydraulic1 component is fixed to the origin.

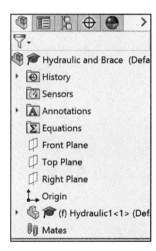

4. **Insert** the BRACE component.

5. **Create** a Concentric mate between the inside cylindrical face of Hydraulic1 and the outside cylindrical face of BRACE.

6. **Create** a Parallel mate between the outside cylindrical face of Hydraulic1 and the outside cylindrical face of BRACE.

7. **Create** a Distance mate (20mm) between the inside bottom face of Hydraulic1 and the bottom flat face of BRACE as illustrated.

8. **Calculate** the mass of the assembly in grams.

9. **Enter 51.55** grams. Always enter the needed decimal places (in this case 2) in the answer field.

10. **Save** the assembly.

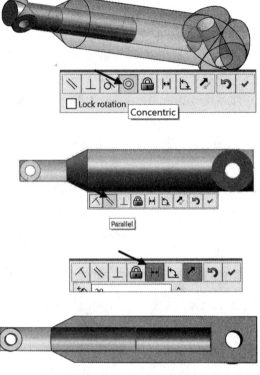

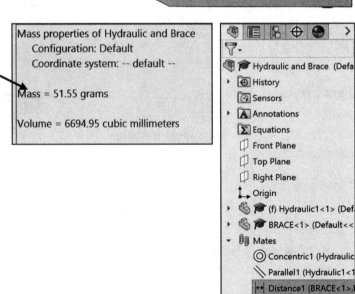

Mass properties of Hydraulic and Brace
    Configuration: Default
    Coordinate system: -- default --

Mass = 51.55 grams

Volume = 6694.95 cubic millimeters

A question in this segment could be - Insert the Hydraulic and Brace assembly into the Welding Arm assembly.

Suppress the Distance mate in the Hydraulic and Brace assembly.

Make the Hydraulic and Brace assembly flexible.

Mate and position the Hydraulic and Brace assembly in the Welding Arm assembly.

Set the proper faces (mates) as indicated in the images.

Insert a Width mate to properly position the Hydraulic and Brace assembly in between the mounting holes.

Insert an Angle mate of 25 degrees between the Holder-Thongs-A assembly and the TopFixture -A assembly.

Calculate the mass in grams of the Welding Arm assembly.

Another question could be - calculate the center of gravity (mm) relative to Coordinate System1.

Provided Information:

Unit system: MMGS (millimeter, gram, second)

Decimal places: 2

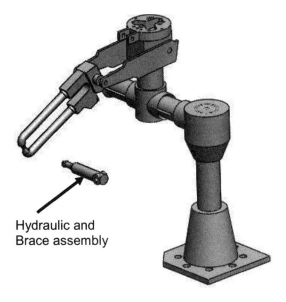

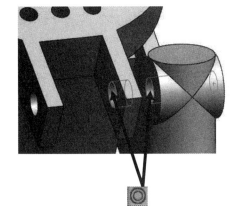

Hydraulic and Brace assembly

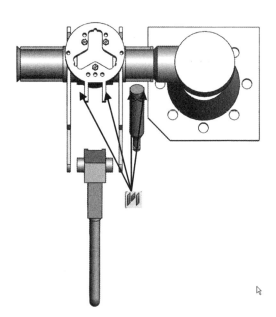

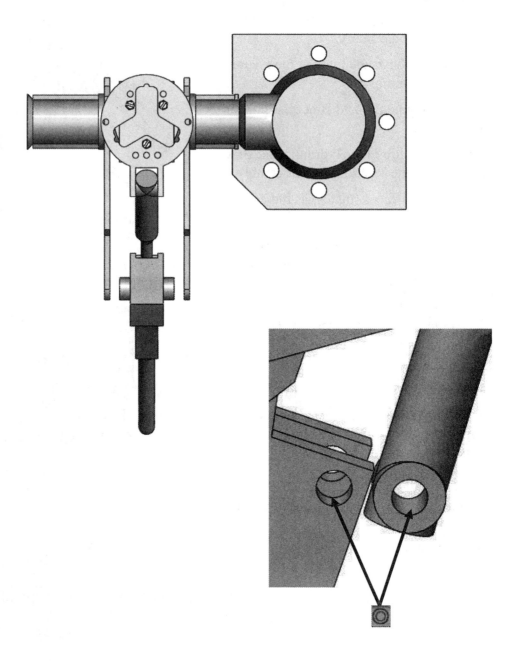

Hydraulic and
Brace assembly

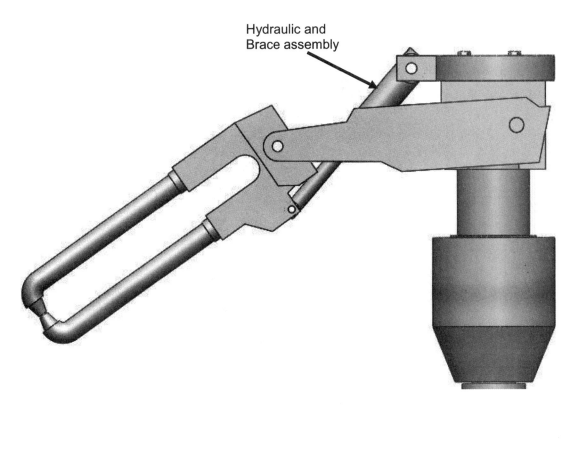

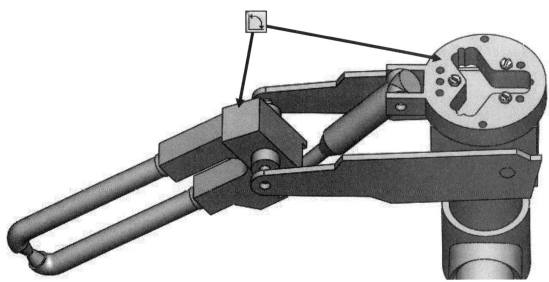

Let's begin.

1. **Insert** the Hydraulic and Brace assembly into the Welding Arm assembly.

2. **Suppress** the Distance mate in the Hydraulic and Brace assembly.

3. **Create** a Concentric mate between the inside cylindrical face of the Hydraulic and Brace assembly and the inside cylindrical face of the Fixture tabs.

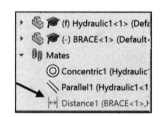

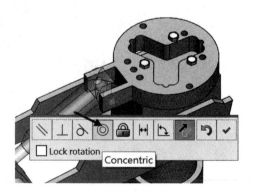

4. **Create** a Width mate between the two tabs (outside faces) of the Fixture and the two outside cylindrical faces of Hydraulic1. The Hydraulic and Brace assembly is free to rotate. When a sub-assembly is inserted into an assembly it is in the rigid state.

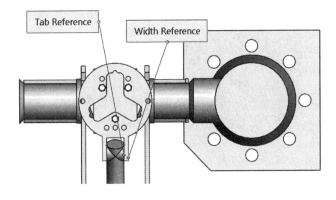

5. **Create** a Flexible state for the Hydraulic and Brace sub-assembly. Note the icon change in the Assembly FeatureManager.

The Hydraulic and Brace assembly is free to translate. Use this ability to translate and to align the end holes of the sub-assembly to the brace of the Welding Arm assembly.

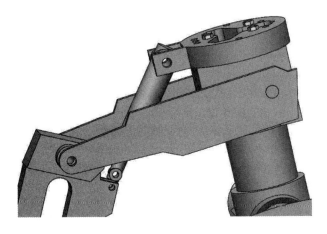

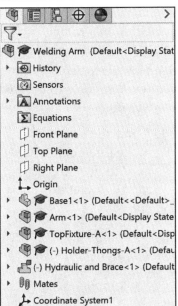

6. **Insert** a Concentric mate between the two cylindrical faces as illustrated.

7. **Insert** a (25 degree) Angle mate between the flat face of the Thong holder and the top face of the Fixture. Note: the angle dimension is referenced from the top of the Fixture.

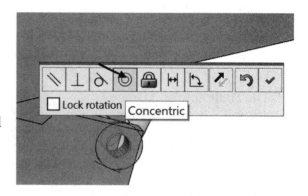

8. **Calculate** the mass of the assembly in grams.

9. **Enter 4609.45** grams. Always enter the needed decimal places in the answer field.

10. **Calculate** the center of gravity (mm) relative to the Coordinate System1.

11. **Enter** the center of gravity (mm) in the three blank fields. You need to be within 1% of the answer to get this question correct.

    X = -92.11

    Y = 40.93

    Z = 134.49

12. **Save** the assembly.

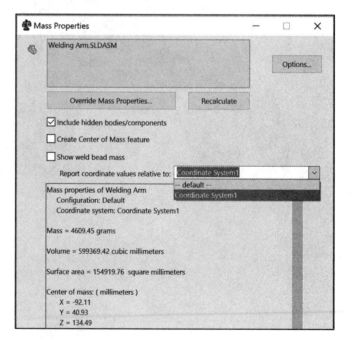

Use the Measure tool to confirm your Angle mate dimension.

During the exam, you will be forced into an error situation.

A question in this segment could be - download and open the attached components (Spring Pin Slotted _AI_1-5, Dowel Pin_AI, and Spring Pin Slotted_AI_5.

Insert the Spring Pin Slotted _AI_1-5 component, the Dowel Pin_AI component, and the Spring Pin Slotted_AI_5 component into the Welding Arm assembly.

Set the proper faces (mates) as indicated in the images.

Insert Concentric and Width mates.

Modify the dimension of the Spring Pin Slotted _AI_1-5 component to be flush with the outside face of the Support Arms (50mm).

Provided Information:

Unit system: MMGS (millimeter, gram, second)

Decimal places: 2

Calculate the mass of the Welding Arm assembly in grams.

Spring Pin Slotted _AI_1-5

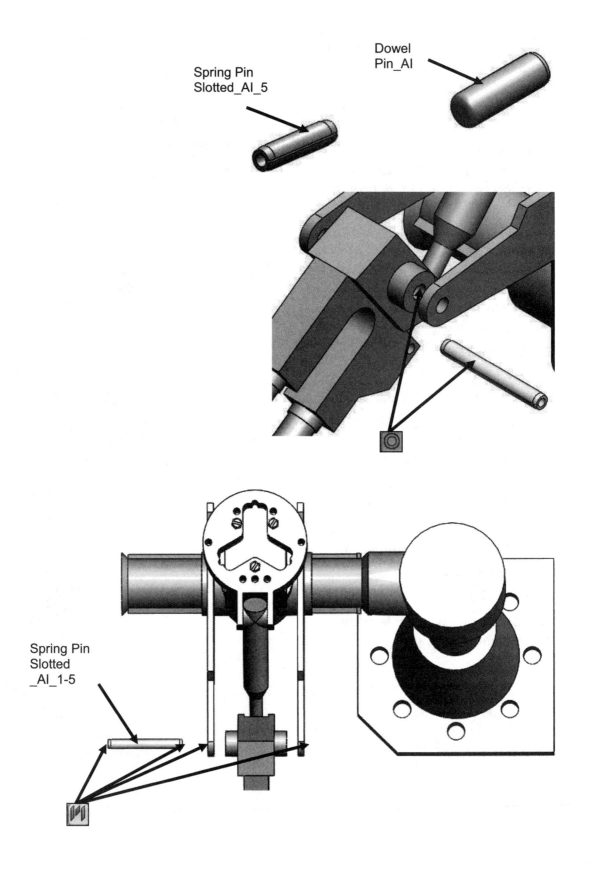

Spring Pin
Slotted_AI_5

Dowel
Pin_AI

Spring Pin
Slotted
_AI_1-5

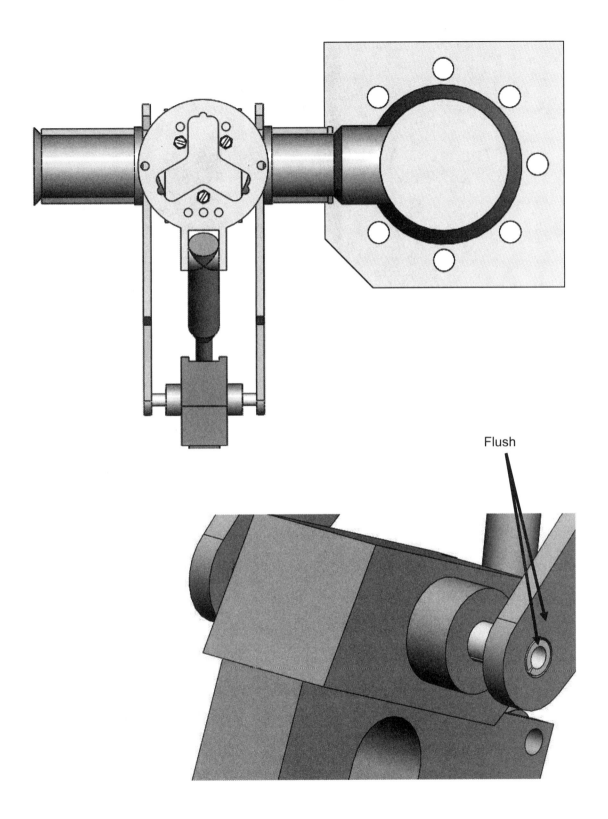

Flush

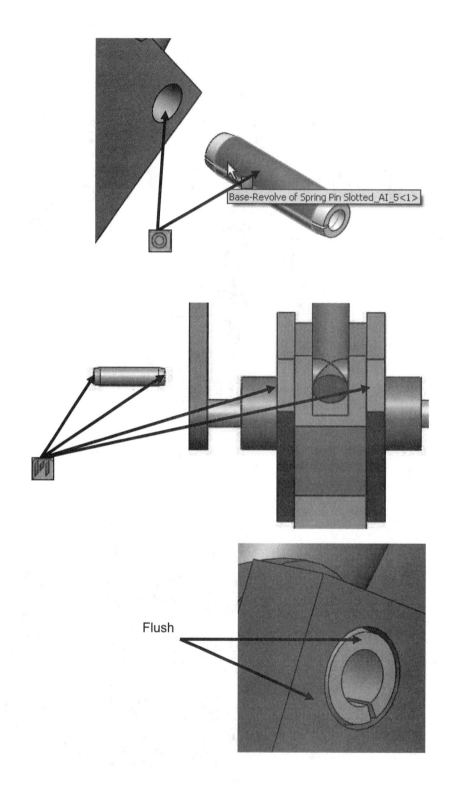

Flush

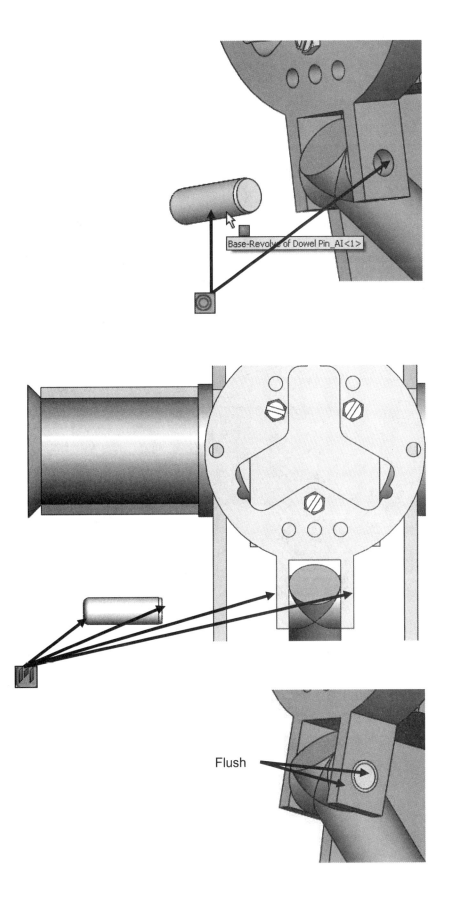

Flush

Let's begin.

1. **Insert** the Spring Pin Slotted _AI_1-5 component, the Dowel Pin_AI component, and the Spring Pin Slotted_AI_5 component (from Segment 3 Initial/Initial 5 - add pins folder) into the Welder Arm assembly.

2. **Create** a Concentric mate between the outside cylindrical face of the Spring Pin Slotted _AI_1-5 component and inside cylindrical face of the TopFixture-A component.

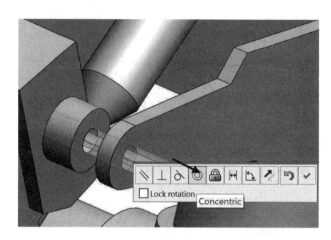

3. **Create** a Width mate between the Spring Pin Slotted _AI_1-5 component and the two support arms of the assembly. Note: the Spring Pin is not flush with the two support arms.

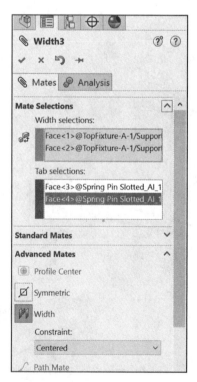

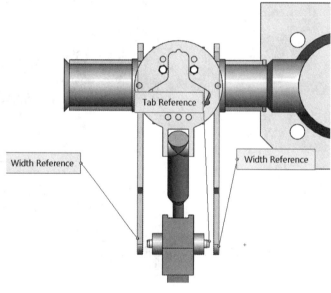

4. **Modify** the Spring Pin Slotted _AI_1-5 component length dimension to 50mm (flush) with the outside face of the two support arms of the assembly. It is important that you know how to modify dimensions of a component inside an assembly for the exam.

💡 Use the Measure tool to confirm the required dimension.

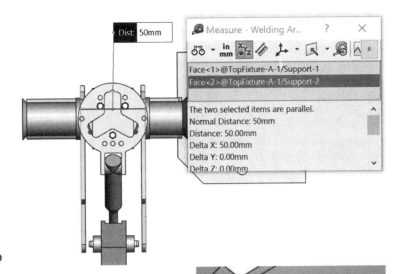

5. **Create** a Concentric mate between the outside cylindrical face of the Spring Pin Slotted_AI_5 component and the inside cylindrical face of the Holder-Thongs-A component.

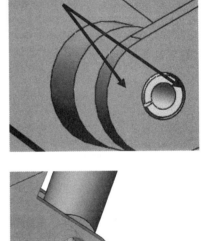

6. **Create** a Width mate between the Spring
Pin Slotted_AI_5 component and the two
faces of the Holder-Thong of the assembly.
The faces are flush.

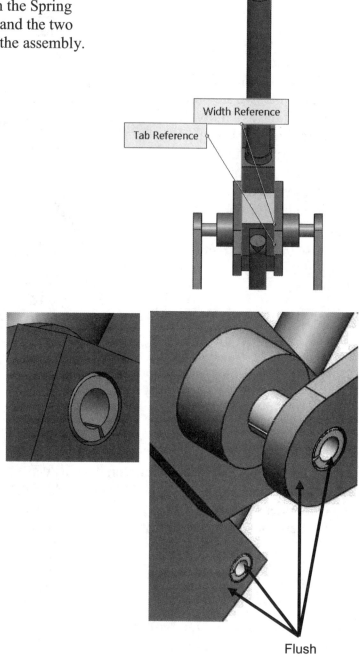

Flush

7. **Create** a Concentric mate between the outside cylindrical face of the Dowel Pin_AI component and the inside cylindrical face of the TopFixture-A component.

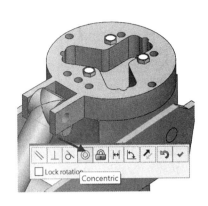

8. **Create** a Width mate between the Dowel Pin_AI component and the two faces. The faces are flush.

9. **Calculate** the mass of the assembly in grams.

10. **Enter 4623.00** grams. Always enter the needed decimal places in the answer field.

Another question could be - calculate the center of gravity (mm) of the Welding Arm assembly relative to Coordinate System1.

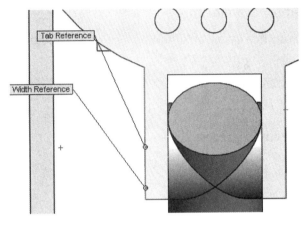

11. **Calculate** the center of gravity (mm) relative to the Coordinate System1. Enter the center of mass in the three blank fields in the exam.

    X = -92.28

    Y = 40.82

    Z = 134.84

12. **Save** the assembly.

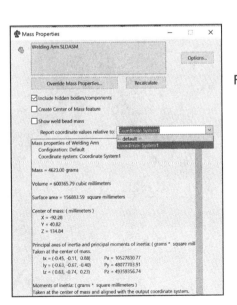

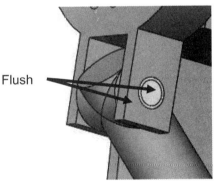

Flush

A question in this segment could be - download and open the attached component (BaseB).

Replace the Base1 component with the BaseB component.

Mate the Arm to BaseB using the same mates as with the Base1 component.

Maintain the same mates of the rest of the components as previously directed.

Redefine Coordinate System1 with BaseB as illustrated.

Using the created coordinate system as the Output Coordinate System, calculate the center of gravity (mm) of the assembly.

Another question could be - calculate the mass of the assembly in grams.

Provided Information:

Unit system: MMGS (millimeter, gram, second)

Decimal places: 2

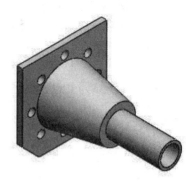

Base1 part

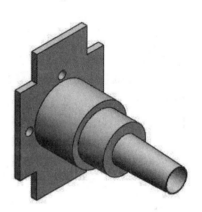

BaseB part

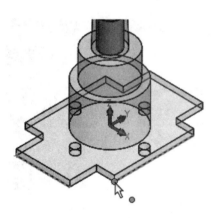

Let's begin.

Replace the Base1 component in the Welding Arm assembly.

1. **Right-click** Base1 in the Assembly FeatureManager.

2. Click **Replace Components**.

3. **Browse** to the location that you downloaded the book models (Segment 3 Initial/Initial 5 - replace base folder).

4. **Double-click** BaseB.

5. Click **OK** from the PropertyManager.

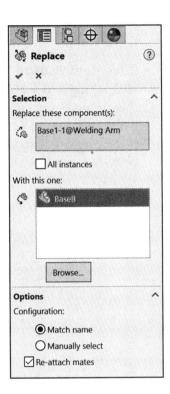

The What's Wrong dialog box is displayed. Isolate the Base component. This section presents a representation of the types of questions that you will see in this segment of the exam. Depending on your selection of faces and order of components inserted into the top level assembly, your error messages may vary. Address errors as needed.

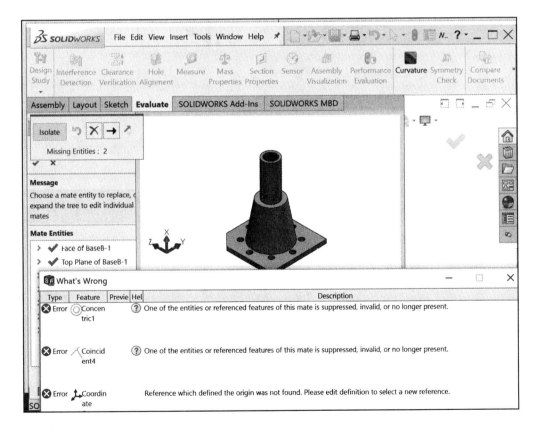

Replace the missing faces to address the mate errors.

6. **Click** Isolate.

7. Click **Replacement Part Only (Selected Entity)**. The Base1 part is displayed in the isolate window. The BaseB part is displayed in the Graphics window.

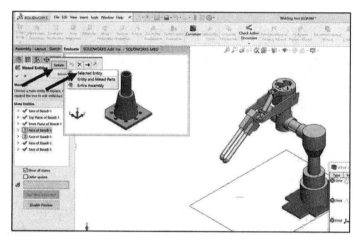

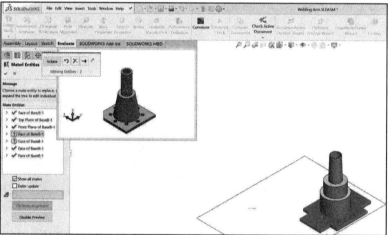

8. **Click** the inside cylindrical face of BaseB as illustrated. The Concentric1 error is removed.

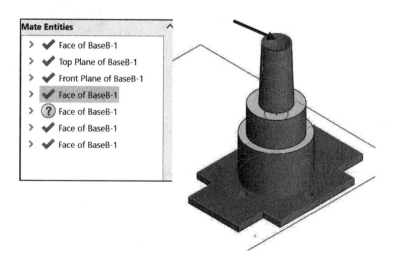

9. **Expand** the Face of BaseB-1 Mate Entities.

10. **Click** Coincident4.

11. **Click** inside the flat bottom circular face of BaseB as illustrated. The error is removed.

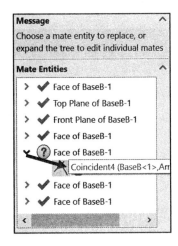

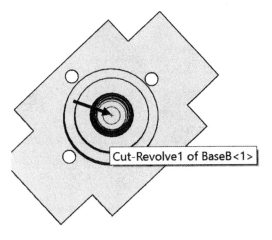

Cut-Revolve1 of BaseB<1>

12. **Click** Entire Assembly from the Isolate box. The arm is in the correct orientation. If not, click the Flip Mate Alignment button.

13. **Click** OK from the Mated Entities PropertyManager. There is still an error with the coordinate system.

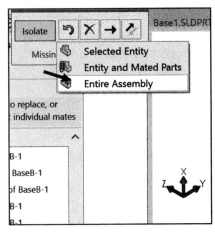

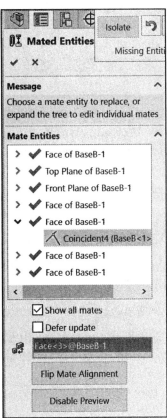

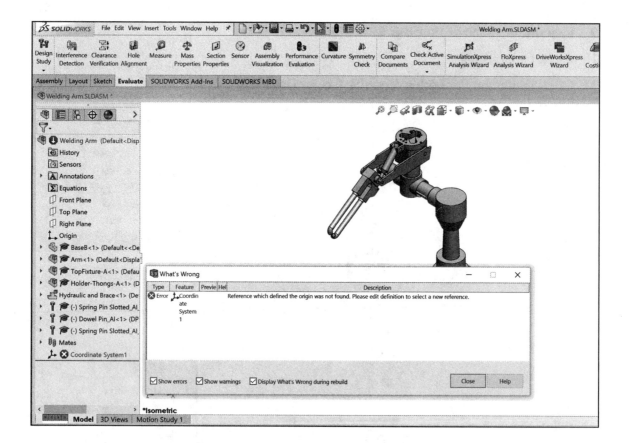

Confirm the position of the Arm component inside the Welding Arm assembly. Use the Change Transparency tool or a section view to confirm the location of mated components in the assembly.

14. **Right-click** Change Transparency from the BaseB component in the Assembly PropertyManager.

15. **Close** all dialog boxes.

Address the error for Coordinate System1. Redefine the vertex.

16. **Edit** Coordinate System1.

17. **Click** the bottom right front vertex on the BaseB component as illustrated.

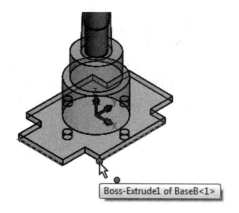

18. **Calculate** the mass of the assembly in grams.

19. **Enter 5251.51** grams. Always enter the needed decimal places in the answer field.

20. **Calculate** the center of gravity (mm) relative to the redefined Coordinate System1. Enter the center of mass in the three blank fields in the exam.

    X = -87.30

    Y = 36.84

    Z = 130.27

21. **Save** the assembly.

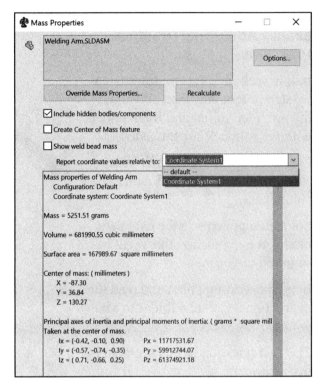

In the exam you may see additional mate errors.

Depending on your selection of faces and order of components inserted into the top level assembly, your error messages may vary.

Segment 3 of the CSWP CORE exam - Additional Practice Problems

In this section, there are fewer step-by-step procedures than above. Use the provided initial and final models with the rollback bar if needed.

Create an assembly (Bench Vise) starting with the Bench Vise-Base component.

Some components in the exam are modeled by you and others will be supplied to you.

Load the Testing client and read the instructions. Create a folder to save your working models.

The first question in this segment is instructional.

The second question can be - download the Bench Vise-Base component and create the Bench Vise assembly.

Insert and orientate the Bench Vise-Base component as illustrated below with the Red tab in the front location.

Provided information:

Units: IPS (inch, pound, second)

Decimal Places: 2

Part Origin: Arbitrary

All holes through all unless shown otherwise.

There are no inferences.

The Top Plane of the assembly is perpendicular with the Front Plane of the Bench Vise-Base component.

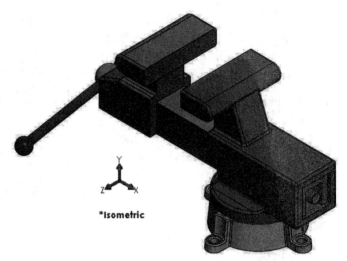

*Isometric

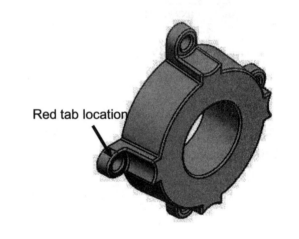

Red tab location

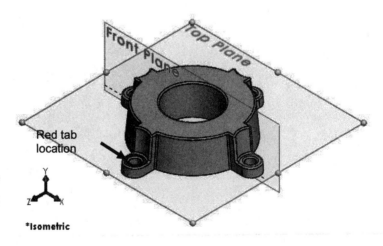

Red tab location

*Isometric

What is the center of gravity of the Bench Vise-Base component in the assembly?

🔆 During the exam, SOLIDWORKS will provide a part to create that is not orientated correctly in the assembly. Knowledge of component orientation in an assembly is required along with creating coordinate systems.

Let's begin.

1. **Create** an assembly document.

2. **Insert** the Bench Vise-Base component (Segment 3 Initial/Bench Vise Parts folder). By default the Bench Vise-Base is fixed to the origin. Set document units and precision.

There are numerous ways to modify the orientation of a component in an assembly document that is fixed to the origin.

Use the Float and Mate tools in the next section. Re-orientate the Bench Vise-Base component. Fully define the component in the Assembly document.

3. **Float** the part.

4. **Re-orientate** the assembly as needed. Create a Coincident mate between the Right Plane of the Bench Vise-Base component and the Top Plane of the assembly.

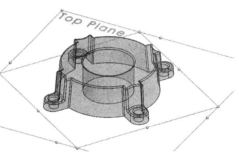

5. **Click** the Aligned option if needed.

Create two additional Coincident mates.

6. **Create** a Coincident mate between the Right Plane of the Assembly and the Top Plane of the Bench Vise-Base component.

Another mate is needed to fully define the component.

7. **Create** a Coincident mate between the Front Plane of the Assembly and the Front Plane of the Bench Vise-Base component. The Bench Vise-Base component is fully defined in the Assembly document.

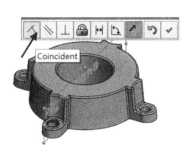

🔆 You can also rotate the Bench Vise-Base component in the assembly using the Rotate Component tool (about Y&X).

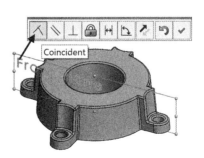

8. **Calculate** the center of gravity in inches of the Bench Vise-Base component in the assembly.

9. **Enter** the center of gravity in the three blank fields (inches). You need to be within 1% of the answer to get this question correct.

   X = 0.00

   Y = 0.59

   Z = 0.00

10. **Save** the assembly. In this example name the assembly Bench Vise.

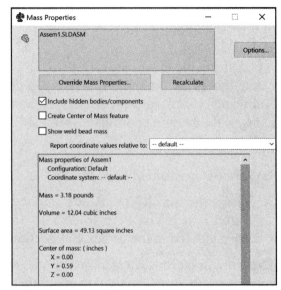

SOLIDWORKS Mass Properties calculates the center of mass for every model. At every instant of time, there is a unique location (x, y, z) in space that is the average position of the systems mass.

There are numerous ways to build the model in this section. A goal is to display different design intents and techniques.

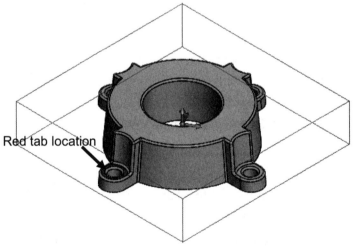

Red tab location

Question 2:

A question in this segment could be - download and open the Bench Vise-Main Body sub assembly. Insert the Bench Vise-Main Body sub assembly into the Bench Vise assembly.

Position the Bench Vise-Main Body assembly with respect to the Bench Vise-Base component as illustrated.

All parts are Coincident and there are no inferences.

Insert all needed mates.

What is the center of gravity of the Bench Vise assembly?

Provided Information:

Unit system: IPS (inch, pound, second)

Decimal places: 2

SOLIDWORKS provides symbols to indicate a Parallel mate, Width mate, Coincident mate, etc. in the exam.

Take your time to review the drawing views and to understand the required mates between each component.

In this example a Parallel mate is needed between the Front Plane of the Bench Vise-Main Body sub assembly and the Front Plane of the Bench Vise assembly.

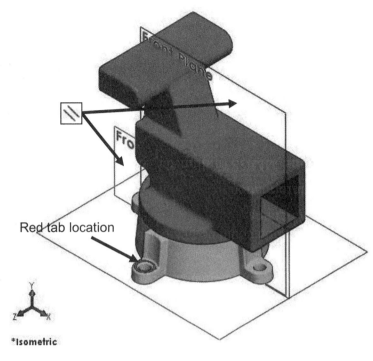

Red tab location

*Isometric

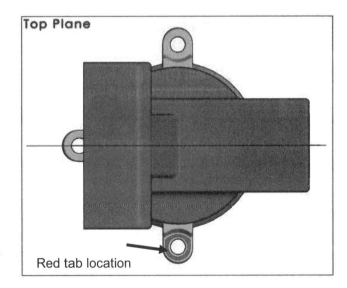

Top Plane

Red tab location

*Top

Let's begin.

Insert the Bench Vise-Main Body sub assembly.

1. **Insert** the Bench Vise-Main Body sub assembly (from the Segment 3 Initial/Bench Vise Parts folder) into the Bench Vise assembly.

2. **Insert** a Coincident mate between the temporary axis of the Bench Vise-Base component and the temporary axis of the Bench Vise-Main Jaw Body sub assembly.

It is good practice to save frequently and to rename the part or assembly if you need to go back during the exam.

3. **Insert** a Coincident mate between the top face of the Bench Vise-Base component and the bottom face of the Bench Vise-Main body. The Bench Vise is free to rotate.

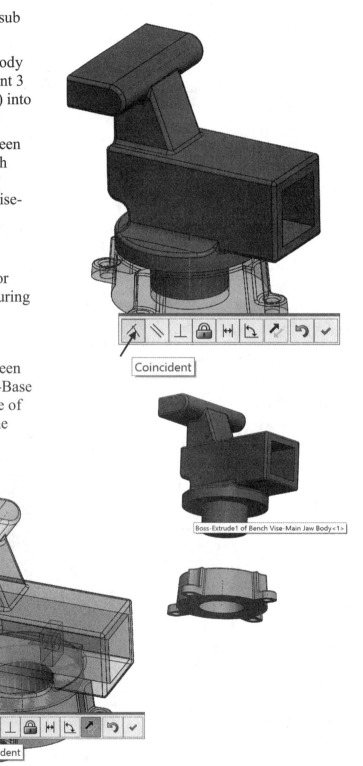

Next, restrict the rotation of the vise.

4. **Insert** a Parallel mate between the Front Plane of the Bench Vise-Main Body sub assembly and the Front Plane of the Bench Vise assembly.

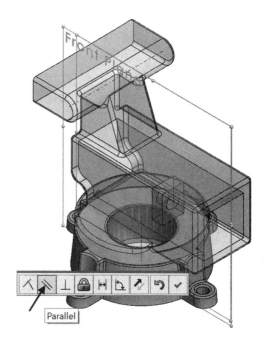

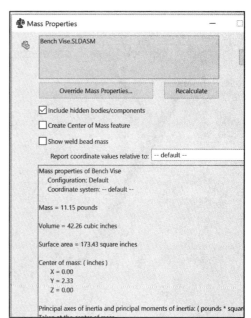

5. **Calculate** the center of gravity of the Bench Vise assembly.

6. **Enter** the center of gravity (inch) in the three blank fields. You need to be within 1% of the answer to get this question correct.

X = -0.00

Y = 2.33

Z = 0.00

7. **Save** the assembly.

💡 You need to be within 1% of the answer to get this question correct.

💡 There are numerous ways to address the question in this section. A goal is to display different design intents and techniques.

Question 3:

A question in this segment could be - download and open the Bench Vise-Sliding Jaw and Bench Vise-Threaded rod component.

Insert the components into the Bench Vise assembly.

Calculate the Normal Distance (X) between the end face of the Bench Vise-Threaded rod to the back face of the Bench Vise-Threaded rod holder.

Insert a Distance mate between the face of the Bench Vise-Sliding Jaw and the face of the Bench Vise-Main Jaw body.

The Front Plane of the assembly is parallel with the Front Plane of the Bench Vise Sliding Jaw component.

The Bench Vise-Threaded rod is free to rotate about its axis.

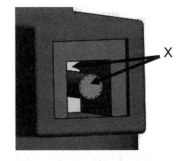

Bench Vise-Threaded rod

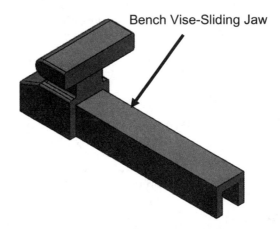

Bench Vise-Sliding Jaw

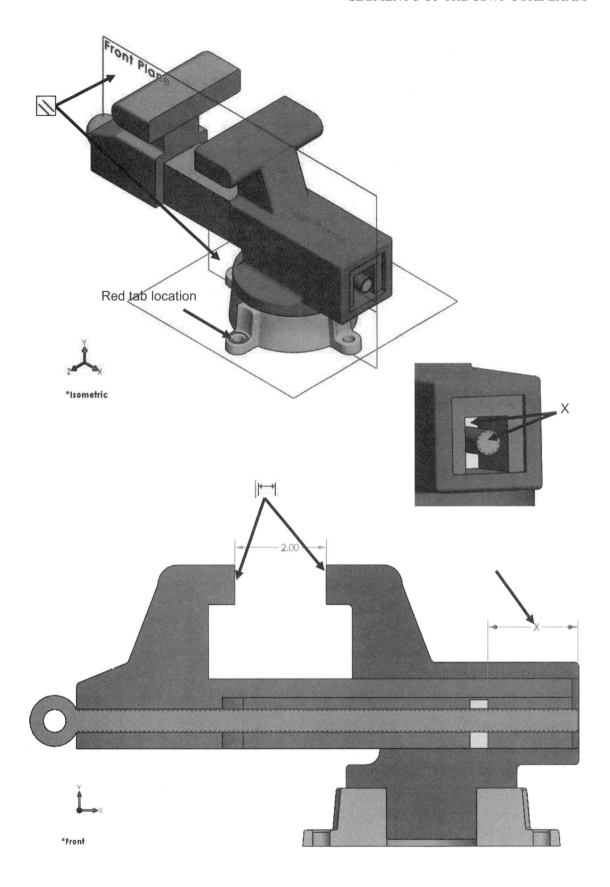

Red tab location

*Isometric

X

2.00

X

*Front

Provided Information:

Unit system: IPS (inch, pound, second)

Decimal places: 2

Let's begin.

1. **Insert** the Bench Vise-Sliding Jaw and the Bench Vise-Threaded rod (from the Segment 3 Initial/Bench Vise Parts folder) into the Bench Vise assembly.

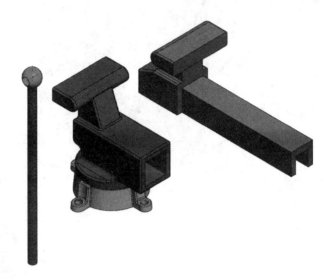

2. **Insert** two Coincident mates to mate the Bench Vise-Sliding Jaw into the Bench Vise-Main jaw body.

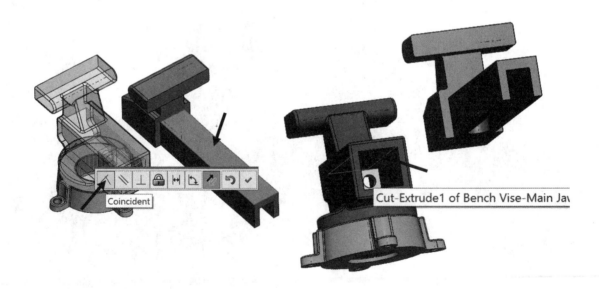

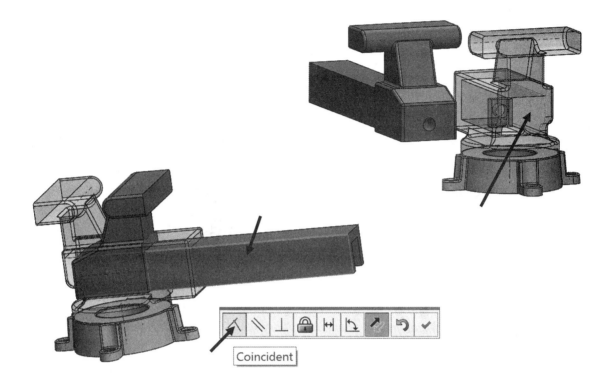

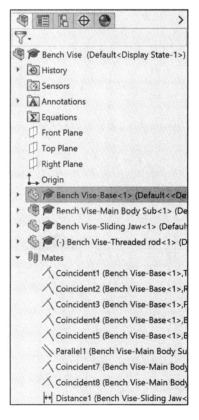

3. **Insert** a Distance mate (2 inches) between the face of the Bench Vise-Sliding Jaw and the face of the Bench Vise-Main Jaw body jaws of the vise.

4. **Save** the assembly.

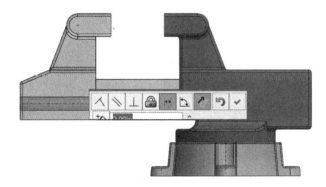

5. **Insert** a Coincident mate between the edge of the Bench Vise-Threaded rod and the hole edge of the Bench Vise-Sliding Jaw component as illustrated. The Bench Vise-Threaded rod is free to rotate about its axis.

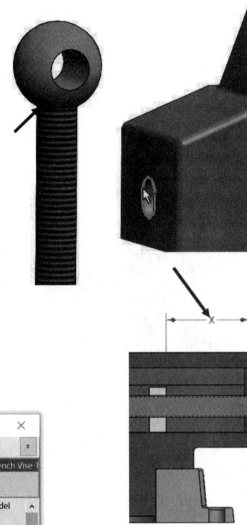

Calculate the Normal Distance of X shown in the image. X is the distance from the end of the threaded rod to the back of the threaded rod holder.

6. **Apply** the Measure tool. Measure the distance between the end of the threaded rod to the back of the threaded rod holder.

7. **Enter 1.98** for the answer to this question. Always enter the needed decimal places in the answer field.

8. **Save** the assembly.

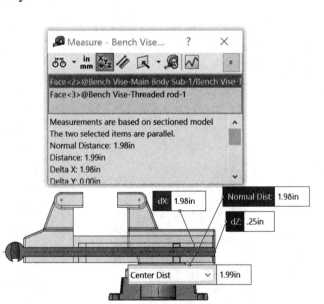

Question 4:

A question in this segment could be - download and open the Bench Vise Handle. Insert the Bench Vise Handle into the assembly. Mate the rod according to the below illustrations.

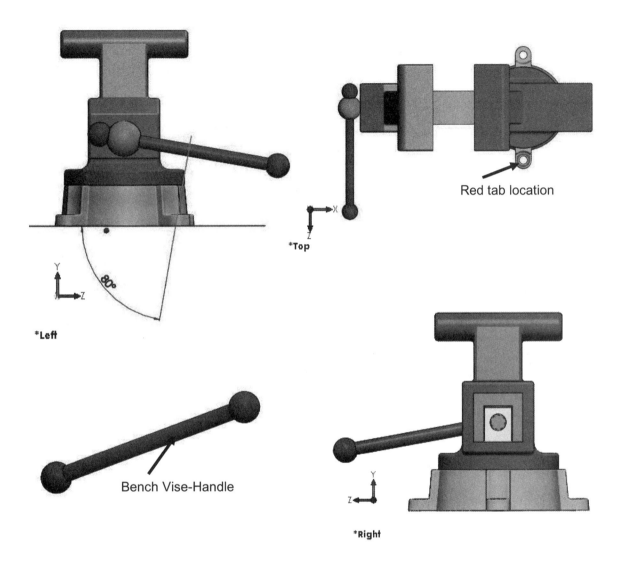

Red tab location

⁺Top

⁺Left

Bench Vise-Handle

⁺Right

To define the axial translation, apply the Move tool with Collision Detection to determine the position. The location is where the ball of the handle is Coincident with the hole in the threaded rod.

🔆 Knowledge of Collision Detection and the Measure tool is required for this section.

What is the center of gravity of the assembly?

Provided Information:

Unit system: IPS (inch, pound, second)

Decimal places: 2

Let's begin.

1. **Insert** the Bench Vise-Handle from the (Segment 3 Initial/Bench Vise Parts folder) into the Bench Vise assembly.

2. **Insert** a Concentric mate between the face of the hole of the threaded rod and the cylindrical face of the handle.

3. **Insert** an 80 degree Angle mate between the Right Plane of the Bench Vise-Handle and the Top Plane of the assembly. Apply the Flip Dimension option if needed.

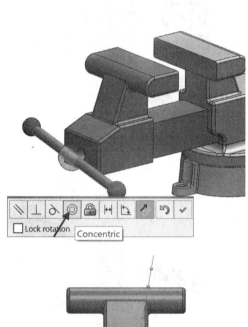

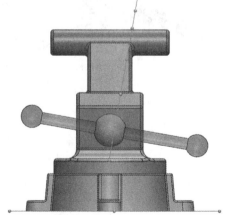

4. **Activate** Collision detection from the Rotate Component PropertyManager. Locate where the Bench Vise-Handle will be coincident with the hole in the Bench Vise-Threaded rod. Note: You can also insert a Distance mate of 0.

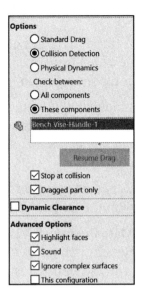

5. **Calculate** the center of gravity of the assembly.

6. **Enter** the center of gravity in the three blank fields. You need to be within 1% of the answer to get this question correct. Always enter the needed decimal places in the answer field.

X = -1.86

Y = 2.85

Z = 0.03

7. **Save** the assembly.

☼ There are numerous ways to build the assembly in this section. A goal is to display different design intents and techniques.

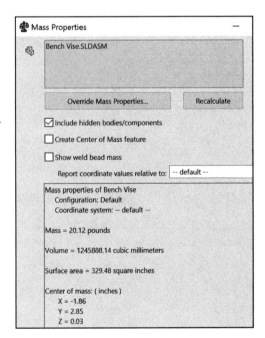

Question 5:

A question in this segment could be - download and open the attached component Bench Vise-Base2.

Calculate the center of gravity in mm of the assembly relative to the location of the red tab location on the Bench Vise Base2 component.

Replace the Bench Vise-Base component with the Bench Vise-Base2 component.

Mate the Bench Vise to the Bench Vise-Base2 component using the same mates as with the Bench Vise-Base.

Maintain the same mates of the rest of the components as previously directed.

There are no interferences in the assembly.

Provided Information:

Unit system: MMGS (millimeter, gram, second)

Decimal places: 2

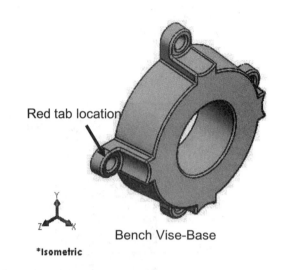

Red tab location

Bench Vise-Base

*Isometric

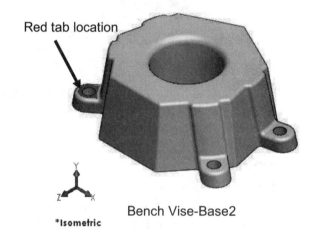

Red tab location

Bench Vise-Base2

*Isometric

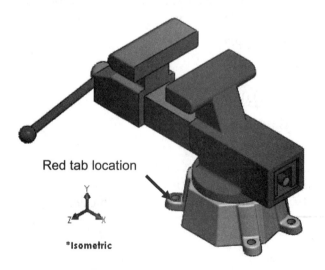

Red tab location

*Isometric

Let's begin.

Replace the Bench Vise-Base component in the Bench Vise assembly.

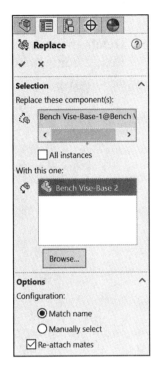

1. **Right-click** Bench Vise-Base in the Assembly FeatureManager. Click Replace Components.

2. **Browse** to the location that you downloaded the models (Segment 3 Initial/Bench Vise Parts folder).

3. **Double-click** Bench Vise-Base1. Click **OK** from the Replace PropertyManager.

Unselect the Re-attach mates box in the Replace PropertyManager to manually address all needed mates.

The What's Wrong dialog box is displayed. Isolate the Base component. This section presents a representation of the types of questions that you will see in this segment of the exam. Depending on your selection of faces and order of components inserted into the top level assembly, your error messages may vary. Address errors as needed.

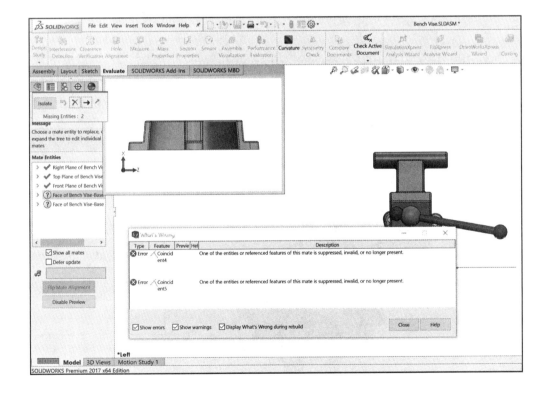

Replace the missing faces to address the mate errors.

4. **Click** Isolate. Select **Entity and Mated Parts**.

5. **Delete** the mate between the Top Plane of the Bench Vise-Base 2 component (Concident2) and the Right Plane of the assembly.

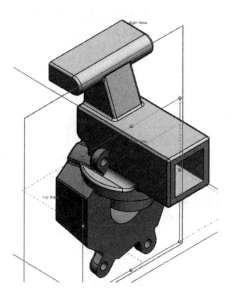

6. **Modify** the Coincident mate that is currently with the Right Plane of the Bench Vise-Base2 and the Top Plane of the assembly to the Top Plane of the Bench Vise-Base2 component with the Top Plane of the assembly.

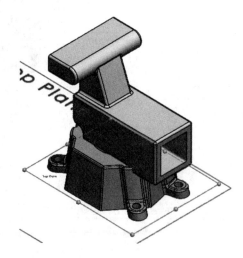

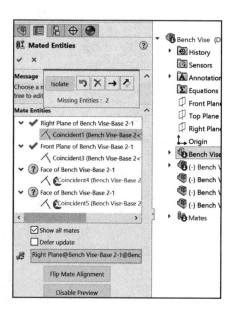

7. **Expand** the second Mate Entity error as illustrated.

8. **Click** Coincident5. It has a missing reference with the top face of Bench Vise-Base. **Select** the Top face of Bench Vise-Base 2.

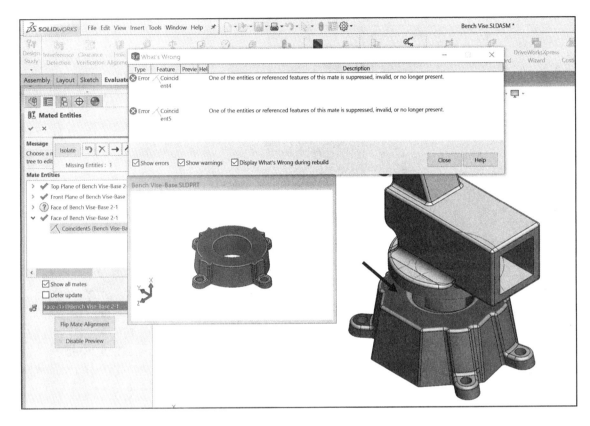

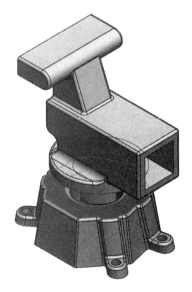

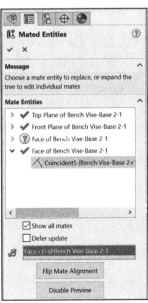

9. **Expand** the first Mate Entity error.

10. **Click** Coincident4. Replace the missing reference.

11. **Select** the temporary axis of Bench Vise-Base 2.

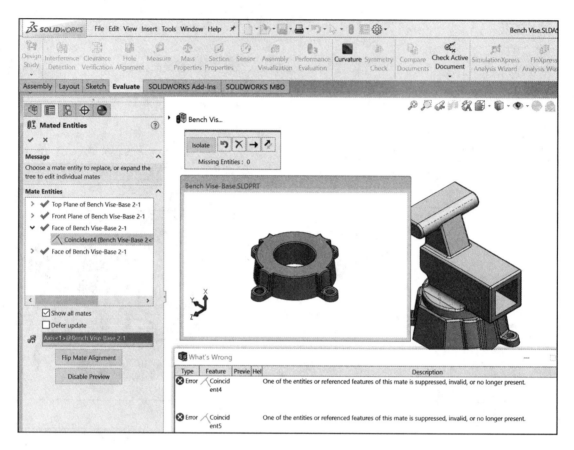

12. **Click** OK from the Mated Entities PropertyManager.

13. **Insert** a Coincident mate between the Right Plane of the Bench Vise-Base2 and the Right Plane of the assembly. The model is fully constrained.

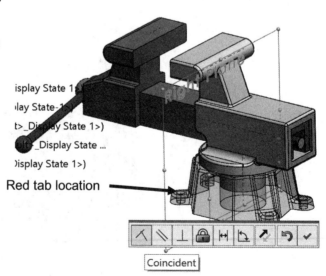

14. **Modify** the assembly documents to MMGS.

15. **Calculate** the center of gravity in mm of the assembly relative to the location of the red tab location on the Bench Vise Base2 component.

16. **Enter** the center of gravity (mm) in the three blank fields. You need to be within 1% of the answer to get this question correct.

    X = -40.13

    Y = 79.94

    Z = 0.67

17. **Save** the assembly. Note: Check the interference between the Bench Vise-Handle (Coincident) and the hole in the Bench Vise-Threaded rod if you are off in the Z direction.

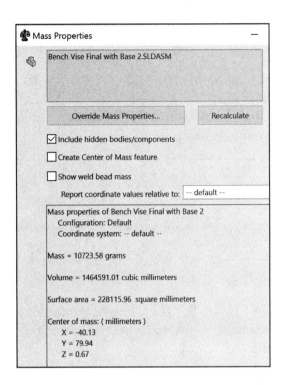

☀ You need to be within 1% of the answer to get this question correct.

☀ There are numerous ways to address the question in this section. A goal is to display different design intents and techniques.

You are finished with this section. Good luck on segment 3 of the CSWP exam.

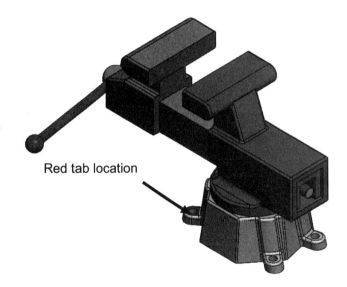

Red tab location

At the end of the exam, view the Summary dialog box. Press the End Examination button only if you are finished.

A total score of 140 out of 190 or better is required to pass the third segment. When you pass the third segment of the CSWP CORE exam you will receive the following email. Click on the SOLIDWORKS Certification Center hyperlink to log in, activate and view your certificate.

Congratulations David!
You have been issued a certificate in SolidWorks® Virtual Test Center indicating that you have successfully completed the requirements for **Segment 3 of Certified SolidWorks Professional Core.**

CertificateID:

Visit your certificate area on SolidWorks® Certification Center. Click My Account to access your certificates and exams.

Congratulations,
Avelino Rochino

If you fail a segment of the exam, you need to wait 14 days until you can retake that segment. In that time, you can take another segment that you might not have taken until then.

Always save your models to verify your results.

Below are former screen shots from a previous CSWP exam in Segment 3.

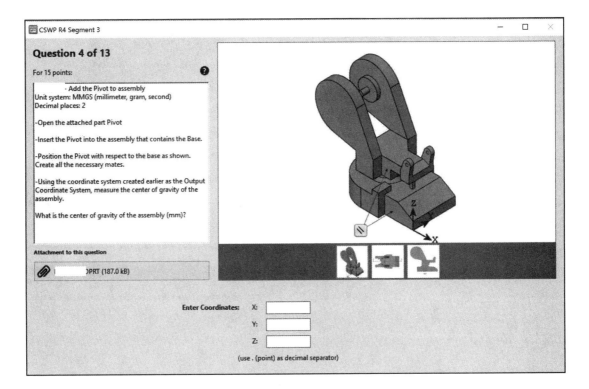

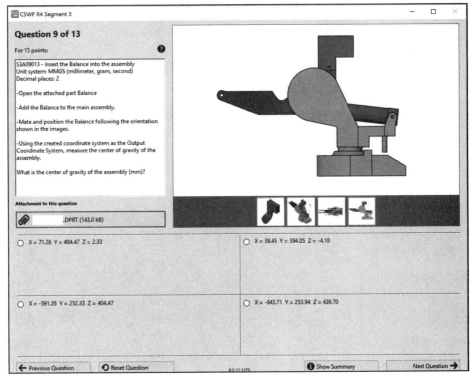

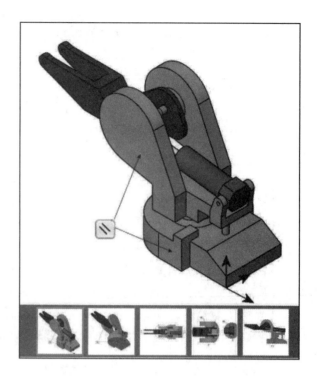

Appendix

SOLIDWORKS Keyboard Shortcuts

Listed below are some of the pre-defined keyboard shortcuts in SOLIDWORKS:

Action:	Key Combination:
Model Views	
Rotate the model horizontally or vertically	**Arrow** keys
Rotate the model horizontally or vertically 90 degrees	**Shift** + **Arrow** keys
Rotate the model clockwise or counterclockwise	**Alt** + left of right **Arrow** keys
Pan the model	**Ctrl** + **Arrow** keys
Magnifying glass	**g**
Zoom in	**Shift** + **z**
Zoom out	**z**
Zoom to fit	**f**
Previous view	**Ctrl** + **Shift** + **z**
View Orientation	
View Orientation menu	**Spacebar**
Front view	**Ctrl** + **1**
Back view	**Ctrl** + **2**
Left view	**Ctrl** + **3**
Right view	**Ctrl** + **4**
Top view	**Ctrl** + **5**
Bottom view	**Ctrl** + **6**
Isometric view	**Ctrl** + **7**
NormalTo view	**Ctrl** + **8**
Selection Filters	
Filter edges	**e**
Filter vertices	**v**
Filter faces	**x**
Toggle Selection Filter toolbar	**F5**
Toggle selection filters on/off	**F6**
File menu items	
New SOLIDWORKS document	**Ctrl** + **n**
Open document	**Ctrl** + **o**
Open From Web Folder	**Ctrl** + **w**
Make Drawing from Part	**Ctrl** + **d**
Make Assembly from Part	**Ctrl** + **a**
Save	**Ctrl** +**s**
Print	**Ctrl** + **p**
Additional shortcuts	
Access online help inside of PropertyManager or dialog box	**F1**

Rename an item in the FeatureManager design tree	**F2**
Rebuild the model	**Ctrl + b**
Force rebuild – Rebuild the model and all its features	**Ctrl + q**
Redraw the screen	**Ctrl + r**
Cycle between open SOLIDWORKS document	**Ctrl + Tab**
Line to arc/arc to line in the Sketch	**a**
Undo	**Ctrl + z**
Redo	**Ctrl + y**
Cut	**Ctrl + x**
Copy	**Ctrl + c**
Additional shortcuts	
Paste	**Ctrl + v**
Delete	**Delete**
Next window	**Ctrl + F6**
Close window	**Ctrl + F4**
View previous tools	**s**
Selects all text inside an Annotations text box	**Ctrl + a**

In a sketch, the **Esc** key un-selects geometry items currently selected in the Properties box and Add Relations box. In the model, the **Esc** key closes the PropertyManager and cancels the selections.

Use the **g** key to activate the Magnifying glass tool. Use the Magnifying glass tool to inspect a model and make selections without changing the overall view.

Use the **s** key to view/access previous command tools in the Graphics window.

SOLIDWORKS Document Types

SOLIDWORKS has three main document file types: Part, Assembly and Drawing, but there are many additional supporting types that you may want to know. Below is a brief list of these supporting file types:

Design Documents	Description
.sldprt	SOLIDWORKS Part document
.slddrw	SOLIDWORKS Drawing document
.sldasm	SOLIDWORKS Assembly document

Templates and Formats	Description
.asmdot	Assembly Template
.asmprp	Assembly Template Custom Properties tab
.drwdot	Drawing Template
.drwprp	Drawing Template Custom Properties tab
.prtdot	Part Template
.prtprp	Part Template Custom Properties tab
.sldbt	General Table Template
.slddrt	Drawing Sheet Template
.sldbombt	Bill of Materials Template (Table-based)
.sldholtbt	Hole Table Template
.xls	Bill of Materials Template (Excel)

Library Files	Description
.sldlfp	Library Part file
.sldblk	Blocks

Other	Description
.sldstd	Drafting standard
.sldmat	Material Database
.sldclr	Color Palette File

Helpful Online Information

The SOLIDWORKS URL:
http://www.SOLIDWORKS.com contains
information on Local Resellers, Solution Partners,
Certifications, SOLIDWORKS users' groups and
more.

Access 3D ContentCentral using the Task Pane to
obtain engineering electronic catalog model and
part information.

Use the SOLIDWORKS Resources tab in the
Task Pane to obtain access to Customer Portals,
Discussion Forums, User Groups, Manufacturers,
Solution Partners, Labs and more.

Helpful on-line SOLIDWORKS information is
available from the following URLs:

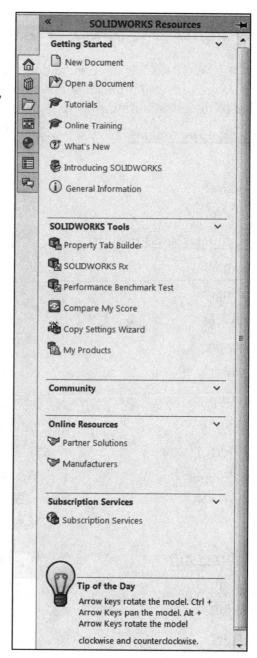

- http://www.mechengineer.com/snug/.

  News group access and local user group
  information.

- http://www.swugn.org/.

  List of all SOLIDWORKS User groups.

- http://www.caddedge.com/SOLIDWORKS-
  user-group-calendar-for-CT-MA-ME-NH-NJ-
  NY-PA-RI-VT.

  Updated SOLIDWORKS information and
  user group calendar for New England and
  surrounding areas.

- http://www.SOLIDWORKS.com/sw/engineer
  ing-education-software.htm.

  SOLIDWORKS in Academia. Information on
  software, support, tutorials, blog and more.

*Online tutorials are for educational purposes only.
Tutorials are copyrighted by their respective owners.

Index